Penguin Education
Modern Science Studies

Modern Physics

Edited by David Webber

Modern Physics

Selected Readings
Edited by David Webber

Penguin Books

Penguin Books Ltd, Harmondsworth,
Middlesex, England
Penguin Books Inc., 7110 Ambassador Road,
Baltimore, Md 21207, USA
Penguin Books Australia Ltd,
Ringwood, Victoria, Australia

First published 1971
This selection copyright © David Webber, 1971
Introduction and notes copyright © David Webber, 1971

Made and printed in Great Britain by
Richard Clay (The Chaucer Press) Ltd, Bungay, Suffolk
Set in Monotype Times

This book is sold subject to the condition that
it shall not, by way of trade or otherwise, be lent,
re-sold, hired out, or otherwise circulated without
the publisher's prior consent in any form of
binding or cover other than that in which it is
published and without a similar condition
including this condition being imposed on the
subsequent purchaser

Contents

Preface 7
SI Units 9

Part One The Quantum Theory 11

1 R. E. Peierls (1964)
The Development of Quantum Theory 17

2 O. R. Frisch (1965)
Take a Photon . . . 33

3 P. Stubbs (1968)
The Introspective Photon 44

4 A. Bairsto (1965)
Lasers 49

Part Two
Nuclear Physics and Fundamental Particles 75

5 R. J. Blin-Stoyle (1959)
The Structure of the Atomic Nucleus 79

6 F. R. Stannard (1966)
*High-Energy Nuclear Physics I
Experimental Techniques* 105

7 F. R. Stannard (1966)
*High-Energy Nuclear Physics II
Fundamental Particles* 120

8 C. Ramm (1968)
High-Energy Neutrinos 135

Part Three The Solid State 155

9 J. M. Ziman (1968)
Solid State 157

10 A. C. Rose-Innes (1965)
The New Superconductors 168

11 A. H. Cottrell (1960)
The Metallic State 191

Part Four **Plasma Physics** 215

12 M. F. Hoyaux (1966)
Plasma Physics and its Applications 217

Part Five **Relativity** 251

13 M. M. Woolfson (1968)
Non-Specialist Relativity 253

14 O. R. Frisch (1961)
Time and Relativity I 281

15 O. R. Frisch (1962)
Time and Relativity II 299

Further Reading 311
Acknowledgements 313
Index 315

Preface

Physics today has become so specialized in its higher reaches that it is difficult for a student in his early studies to get an impression of the scope of modern physics. On the other hand the student who is going to study physics at a more advanced level will want to know what lies beyond his present studies, and it is to such readers that this book is addressed. In making the selection of articles I have looked for those which give an overall picture of certain topics and which have the minimum of difficult mathematics. Consequently the reader will get no more than a glimpse of modern physics but I hope this will be enough to encourage him to follow up some of the topics by referring to the books which I have listed as being suitable for further reading.

A word about the terminology often used. The phrase 'classical physics' is used to refer mainly to the theories of physics which had been developed by about 1890: the mechanics of Newton, Lagrange and Hamilton; the laws of electricity, culminating in the theory of electromagnetic waves by Maxwell and Lorentz; the establishment of the wave theory of light, the kinetic theory of heat and the laws of thermodynamics. 'Modern physics' refers to developments which revolutionized many of the concepts of classical physics: the nuclear theory of the atom by Rutherford and Bohr; the quantum theory, beginning with the postulates of Planck about the emission and absorption of radiation; the later developments of these ideas leading to the wave mechanics of Schrödinger, de Broglie and Heisenberg; the theory of relativity developed chiefly by Einstein.

I am well aware that many important topics have been omitted from this selection and I simply plead lack of space. I have tried to include those which seemed to me to involve basic modern physics and to be interrelated in some way. I hope they give some insight into the subject and, above all, encouragement to follow through some of the topics in greater depth.

SI Units

International agreement has been reached between the standards authorities of most of the leading scientific countries to adopt a common system of metric units known as the SI system (Système Internationale d'Unités) and SI units will be increasingly used in universities, colleges and schools in the next few years.

There are six basic SI units:

Basic physical quantity	Name of unit	Symbol
length	metre	m
mass	kilogramme	kg
time	second	s
electric current	ampere	A
thermodynamic temperature	kelvin	K
luminous intensity	candela	cd

In addition there are *derived units* which are defined in terms of the basic units or other derived units, including the following:

Physical quantity	Name of unit	Symbol and definition
force	newton	N (kg m s^{-2})
energy, heat	joule	J (kg m^2 s^{-2})
power	watt	W (kg m^2 s^{-3} = J s^{-1})
electric charge	coulomb	C (A s)
electric potential difference	volt	V (kg m^2 s^{-3} A^{-1} = J A^{-1} s^{-1})
electric resistance	ohm	Ω (kg m^2 s^{-3} A^{-2} = V A^{-1})
frequency	hertz	Hz (s^{-1})
magnetic flux	weber	Wb (kg m^2 s^{-2} A^{-1} = V s)
magnetic flux density	tesla	T (kg s^{-2} A^{-1} = Wb m^{-2})

Agreed prefixes and symbols are used to indicate multiples of units in powers of ten, including the following.

Multiple	Prefix	Symbol
10^6	mega	M
10^3	kilo	k
10^{-3}	milli	m
10^{-6}	micro	μ
10^{-9}	nano	n
10^{-12}	pico	p

The values of some physical constants in SI units and conversion factors are given in the following tables.

Quantity	Value
velocity of light c	$2 \cdot 998 \times 10^8$ m s^{-1}
mass of proton	$1 \cdot 672 \times 10^{-27}$ kg
mass of neutron	$1 \cdot 675 \times 10^{-27}$ kg
mass of electron	$9 \cdot 109 \times 10^{-31}$ kg
charge of electron e	$1 \cdot 602 \times 10^{-19}$ C
Planck constant h	$6 \cdot 626 \times 10^{-34}$ J s

1 C = 3×10^9 e.s.u.
1 Å = 10^{-10} m = 10^{-1} nm = 10^2 pm.
1 gauss = 10^{-4} T.
1 oersted = $(10^3/4\pi)$ A m^{-1}.

Where possible I have changed the units to conform to the SI system, and I am grateful to the authors concerned for permission to do this. Where it has not been convenient to alter the units the above conversion factors are needed.

References

The International System (SI) Units (BS 3763), British Standards Institution, 1964.
The Use of SI Units (PD 5686), British Standards Institution, 1967 (revised edn, 1969).

Part One The Quantum Theory

The paper by Peierls traces the development of quantum theory from its origins to the formulation of the wave equation by Schrödinger, de Broglie and Heisenberg. As this is a review article it assumes a knowledge of the background experimental work, some of which may not be familiar to the reader, so it seems appropriate that we should first have a look at this.

The quantum theory of radiation was first suggested by Planck at the beginning of this century. He was working on the theoretical explanation of the black-body radiation curves, that is graphs which show how the intensity of the radiation from a black body is distributed amongst the wavelengths of the spectrum. The experimental work had led to certain empirical laws, notably those due to Wien, but no successful theoretical derivation of these laws had been achieved. Planck saw that in order to derive an equation which fitted the experimental results, it was necessary to make a new and fundamental assumption about the emission of the radiation. This was that the radiation could be emitted only in packets of energy, and that the energy contained in each packet is proportional to the frequency of the radiation. The constant of proportionality is now called Planck's constant and is denoted by h; it has the value $6·63 \times 10^{-34}$ J s. Thus radiation of frequency v is emitted in packets (called quanta) each of energy hv.

It was later argued by Einstein that not only did emission and absorption involve quanta but that the radiation itself travelled in the same form. These quanta of radiation in transit are called photons. Einstein used the photon theory to explain the photoelectric effect, showing that the maximum energy of the emitted electrons is given by

$$(\tfrac{1}{2}mv^2)_{\max} = h\nu - e\phi,$$

where ϕ is a constant which is different for each metal. Another phenomenon which was explained by the photon theory was the Compton effect. This effect is that when X-rays are scattered by weakly bound electrons, the wavelength of the scattered X-rays is different from that of the incident ones, the extent of the difference being dependent on the angle of scattering. This is unintelligible on a wave picture, which would predict no change of wavelength on scattering. It can however be explained quantitatively by the usual particle collision laws if one takes the X-rays to consist of photons of energy $h\nu$ and momentum $h\nu/c$, where c is the velocity of light.

The next application of the new theory was to the structure of atoms, and in particular to explain their line emission spectra. The atoms of an element, when suitably excited, emit a line spectrum which is characteristic of that particular element. The simplest to investigate is that of hydrogen and much was already known about its spectrum. It was known that the spectral lines formed series, the wavelengths of whose members could be represented by a simple formula. The first of these, due to Balmer, may be written as

$$\frac{1}{\lambda} = R\left[\frac{1}{4} - \frac{1}{m^2}\right],$$

and applies to lines in the visible part of the hydrogen spectrum. The wavelengths λ of each member of the series are given by integral values of m. The constant R is known as the Rydberg constant. A similar equation can be written for another hydrogen series, this one occuring in the infrared region, i.e.

$$\frac{1}{\lambda} = R\left[\frac{1}{9} - \frac{1}{m^2}\right].$$

There are more of these series in the hydrogen spectrum and they can all be summarized in the formula

$$\frac{1}{\lambda} = R\left[\frac{1}{n^2} - \frac{1}{m^2}\right],$$

where the integral value of n determines the series (e.g. $n = 2$ for the Balmer series) and the integral values of m give the individual lines in the series. Now $c = \nu\lambda$, and substituting this in the above equation gives

$$\nu = Rc\left[\frac{1}{n^2} - \frac{1}{m^2}\right].$$

If we apply the ideas of the quantum theory to this equation, we can write

$$h\nu = Rch\left[\frac{1}{n^2} - \frac{1}{m^2}\right]$$

for the energy of the emitted photons. This suggests that the photon energy has come from an energy change of the orbital electron in the hydrogen atom, and as this energy change appears as the difference of two terms, i.e.

$$\frac{Rch}{n^2} - \frac{Rch}{m^2},$$

it suggests that this electron can have only certain discrete energies, and that a photon is emitted when the electron moves from one of these energy levels to another. This idea was developed by Bohr in 1913 and later by Sommerfeld.

Direct experimental evidence for the existence of energy levels in atoms was obtained by Franck and Hertz in 1914. Their experiment involved exciting the electrons in the atoms of a gas by collisions with moving free electrons which had been accelerated through a known potential difference. They found that no excitation was produced unless the free electrons had a certain minimum kinetic energy – the energy needed to excite the bound electrons from one energy level to the next.

The next problem which had to be resolved was the occurrence of doublets (i.e. two lines close together) in the alkali spectra. The best known of these doublets is the one in the orange–yellow region in the spectrum from sodium: the so-called D-lines. This suggests the splitting of the energy levels into two closely spaced levels and this led, as is explained in the article, to the concept of electron spin. Further evidence for this came from a study of the spectra emitted when the excited atoms are subject to a magnetic field. It is found that this causes a splitting of the spectral line; an effect known as the Zeeman effect.

Despite the successes of the theory outlined above there were situations in which it did not lead to correct predictions, and it needed the development of quantum mechanics (or wave mechanics) to complete the theory. The experimental basis for the fundamental ideas of wave mechanics came from the demonstration of electron diffraction, first by Davisson and Germer using reflection of electrons from a metal surface, and then by G. P. Thomson who used thin metal foils through which the electrons could pass. In both cases it was shown that the electrons were not reflected or transmitted uniformly but that there were some directions in which large numbers of electrons travelled and others in which there were very few. A pattern was found, similar to the diffraction pattern formed when X-rays are passed through a crystal. It seems then as if electrons behave like waves as well as like particles. From measurements on the diffraction pattern it was possible to calculate the wavelength of these waves. De Broglie had already predicted the existence of these waves, and had shown theoretically that their wavelength should be h/p, where p is the momentum of the electrons. The experimental measurements confirmed this prediction.

Schrödinger showed how these ideas should replace classical mechanics when making predictions about, for example, the behaviour of the electron in a hydrogen atom. The electron has to be treated as a wave of wavelength h/p, existing in the region surrounding the nucleus. This leads to an equation for the amplitude of this wave in terms of the kinetic energy of the electron and its potential energy in the electric field of the nucleus (which, of course, changes with distance from the nucleus). Schrödinger was able to show that this equation is capable of solution only for certain energies of the electron. It turns out that these energies are precisely the same as the ones used by Bohr in his description of the hydrogen atom, based on the old quantum theory. Thus the experimental evidence from spectra which supported his theory also fits the wave-mechanical predictions. More important, the newer approach can go on to explain experimental observations and make predictions in areas where the old quantum theory was inadequate. It must be emphasized that the old theory is completely replaced by wave mechanics, so that the earlier sections of the article are serving to show how the theories developed and are not presenting modern ideas. Nevertheless it makes fascinating reading to see how the developments took place, and you will find plenty about this in the books on atomic physics and quantum theory listed in the Further Reading list at the end of this book.

The way of interpreting the wave amplitudes obtained from solutions of the Schrödinger equation was indicated by Born. This is to take the square of the amplitude (i.e. the intensity) of the wave in some region as giving the probability of finding the electron in that region. Much use is made of these probability distributions in the explanation of chemical bonding.

At the end of the article there is a reference to a second article by Peierls. This is not reprinted here as much of it is too difficult to be useful at this stage. The article can be found in *Contemporary Physics*, vol. 6, no. 3.

Take a Photon by Frisch highlights one of the conceptual or philosophical difficulties of quantum mechanics, and is followed by a brief description of a recent experiment on the same problem – *The Introspective Photon*.

The final paper, *Lasers* by Bairsto, is included in this section because it describes an interesting and important application resulting from the wave-mechanical interpretation of atomic spectra. The laser is a relatively new invention (1960) and uses for it are still being suggested and investigated. There have been many new developments since this paper was written (1965). Most of the applications mentioned by Bairsto have been more fully explored and in addition there have been new ones. The most notable of these is a process known as holography. Holography is essentially a method of producing a photographic record of a three-dimensional object in such a way that when the photograph (called a hologram) is suitably illuminated and viewed the original object is seen in three dimensions; if there was more than one object in the original scene these are seen to exhibit parallax as the head is moved from side to side, as they would have done in the original. The hologram is made by illuminating the objects with laser light and picking up the reflected radiation on a photographic plate. The plate is simultaneously illuminated with light from the same laser that has been reflected on to it by a mirror. These two sets of waves produce interference patterns where they overlap and these are recorded on the photographic plate to produce the hologram. The hologram consists therefore of interference fringes, and when viewed in ordinary light bears no resemblance to the original object. If, however, the hologram is illuminated from behind, by laser light of the same wavelength which was used to form it, then, on looking through from the front, the original scene is visible as a three-dimensional reconstruction. Details of this and other applications of lasers will be found in the book *Lasers* which is listed in the Further Reading list.

1 R. E. Peierls

The Development of Quantum Theory

R. E. Peierls, 'The development of quantum theory, Part I: Formulation and interpretation', *Contemporary Physics*, vol. 6, 1964, no. 2, pp. 129–39.

Introduction

This article will discuss the growth of the important concepts of quantum theory up to the present and will attempt to indicate where it is obvious that further ideas or further developments are still required.

No attempt will be made to cover the history of the subject fully or to do justice to the contributions by individual authors. Equally, it will be impossible within the available space to describe adequately the important applications of quantum theory. Even from a fundamental point of view, applications are of vital importance because our conviction of the adequacy of the principles comes from their success in the applications. Any applications mentioned are chosen to illustrate points of principle rather than because of their practical importance.

The quantum of action

Looking back at the developments of the beginning of the century one can easily over-simplify the situation as being that classical physics had clearly failed, and that Planck's discovery was the result of dealing with that failure.

In fact, the difficulties of classical theory were clearly established only in the course of the same investigations which already brought out the revolutionary new postulate.

We know today that, amongst the phenomena which classical theory was unable to account for, the most prominent ones are the stability of atoms, and the properties of black-body radiation. The difficulty about the atoms could be seen clearly only when the structure of the atom was understood, i.e. after the work of Rutherford. Until then it had seemed possible that the charged particles inside the atoms, and in particular the electrons, were sitting in, or near, positions of equilibrium in the local field due to the other charges. Even then there would be difficulties, as Planck remarks in one of his papers, because the kinetic theory would require the electrons to contribute to the specific heat, and measurements of the specific heat of gases had shown that this could be fully accounted for by the motion of the atoms, without internal degrees of freedom.

The challenge which led to the new ideas came from the study of radiation in thermal equilibrium. At this time the laws of statistical mechanics were still developing, and we had only just reached the stage where convincing arguments could be made on purely theoretical grounds. At the same time, experiments had led to the knowledge that at any temperature there was a unique energy distribution for radiation in equilibrium, independent of the experimental circumstances.

It had been found to obey Wien's 'Displacement Law', and also Wien's formula, giving an exponential variation of intensity with frequency, which first was believed to be generally valid, and only later found to hold at each temperature only for high enough frequencies. Planck took up the challenge to account for these results, and he first attempted to show that a law of the type of Wien's could be derived from the known principles of the propagation of light, and of statistical mechanics. He convinced himself that the calculations he was using were not compatible with the accepted laws, and that a new hypothesis was required. He had no illusions about the revolutionary nature of the step he was taking, though he emphasized that one must attempt to keep down the degree of change, and make no more drastic assumptions than absolutely necessary.

As we know, the new step was the idea that radiant energy of frequency v could be emitted or absorbed only in amounts of hv.

One could think of this as a characteristic of radiation or of the radiating system, or of the emission and absorption process; in order to minimize the degree of innovation Planck was inclined to associate the new feature with the emission and absorption process. He was then thinking of the emission by harmonic oscillators because one could not otherwise account for the sharp line spectra of atoms; it was therefore necessary to assume that an oscillator of frequency v could lose or gain energy only in multiples of hv. If one wants to ensure that the Planck distribution of radiative energy is maintained even in the presence of other radiating systems, the same restriction would have to apply to any emission or absorption process; if one goes that far one really includes all consequences of the more extreme form of the quantum hypothesis. In particular one then had to reconcile the results of these postulates with Maxwell's laws for the forces on charged bodies in electromagnetic fields, which represent a continuous process, and therefore should allow continuous transfer of energy.

During the years following Planck's first work on the subject much thought was devoted to the question how to reconcile the quantum hypothesis with the apparently established laws of physics. Planck himself, from the desire not to make more revolutionary assumptions than necessary, was inclined to regard the new feature as a property of the emission and absorption process, whereas Einstein argued that for consistency it was necessary to assume the concentration of the energy into discrete quanta, even for a pure radiation field in the absence of matter which interacted with it. He deduced this from arguments about the probability distribution for the energy derived from the determination of the entropy using Planck's radiation formula.

A further important step was taken by Einstein in 1905 when he showed that the photon hypothesis would predict that in the photoelectric effect the energy of the electrons should be independent of the light intensity and a linear function of the light frequency, in good agreement with the then new evidence from the experiment.

Later Einstein pointed out that a detailed consideration of the equilibrium between matter and radiation required light quanta to possess not only energy hv but also a momentum hv/c. This idea, which represented a further step towards giving light quanta the nature of particles, and which prepared the way for the explanation of the Compton effect, was not accepted without reservations by Planck, who pointed out that the argument had to rely on calculations about the exchange of energy between particles and radiation which was not well understood at the time. In the same discussion Planck pointed out that the assumption of light quanta behaving so completely as particles would contradict all the evidence for the wave nature of light, in particular the evidence from diffraction and interference experiments.

It is interesting that Einstein's first reaction to this remark was to express the hope that these phenomena might still be accounted for if one was dealing with a large number of light quanta, in the same way in which the flow of many electrons or other charged particles would lead to behaviour in agreement with a model based on a continuous fluid of electricity.

This hope, of course, was not borne out, both because diffraction phenomena can be observed in circumstances in which only a few light quanta are present in the apparatus at any one time, and also because any effect depending on the mutual interaction between them would lead to consequences depending on the light intensity.

At this time, therefore, the fundamental conflict between the wave and particle descriptions of light was clearly formulated. Already then the idea seems to have been mentioned that the wave field might determine the probability of particles appearing in various regions of space. This idea of waves being a 'guiding field', or 'ghost waves', is usually attributed to Einstein.† It was, however, not until the late 1920s that this idea could be shown to form the basis of a consistent way of reconciling the concept of waves and particles.

The quantum theory of the atom

The work of Rutherford had shown the structure of the atom as consisting of a very small, heavy and positively charged nucleus surrounded by electrons moving around it like planets around the sun. This new knowledge helped to make many facts about atoms intelligible, but it immediately brought new difficulties. Prominent among these was the existence of definite spectral lines, since classical physics would predict a completely continuous spectrum of radiation for a model of the Rutherford

---

† I have been unable to trace any such remark in the literature. Niels Bohr refers to a discussion with Einstein in which he comments on the idea of 'ghost waves' as rather unsatisfactory.

type. Connected with this was the very fact of the stability of atoms, since one would expect an electron in a Kepler type of orbit to go on losing energy by radiation until it would fall into the nucleus. Even apart from radiative processes a similar collapse would result from the mutual interaction of different atoms. Because of the greater complexity of the mechanical problem this difficulty was at the time emphasized less than the problem of radiation.

If one draws an analogy between the atom and the solar system, one must, of course, remember that the size of any planetary orbit is determined only by the previous history of the solar system, and that the reason for its remaining unchanged over long periods is simply the fact that there are no strong radiation effects and no strong mechanical interactions with outside bodies to cause any rapid changes.

Clearly the simplest example on which to test our understanding of the problem was the atom of hydrogen. Balmer's empirical formula for the frequencies of the hydrogen lines, together with the general spectroscopic law of Rydberg by which the frequency of each spectral line could be represented as a difference of two 'terms', formed a clear challenge for the theory.

New deep insight came from the work of Niels Bohr in 1913, who deduced from the facts about atoms the need for a further drastic revision in our thinking about mechanics. This included, firstly, the idea that of the possible orbits predicted by the equations of motion for an electron in the field of the hydrogen nucleus only a discrete set were in fact possible, and secondly, that in the emission or absorption of light by an atom the frequency of the light had to be such as to make the energy $h\nu$ of the light quantum equal to the energy difference of the two electron orbits.

Bohr found that to obtain agreement with the Balmer formula the angular momentum, at least in the case of a circular orbit, had to be a multiple of $h\nu/2\pi$. Here Planck's constant was seen to play a part in the determination of atomic orbits. The angular momentum multiplied by 2π was later seen to be a special case of the 'action integral'

$$\int p \, dq,$$

where q is some coordinate (e.g. the angle) and p the momentum variable conjugate to it (e.g. the angular momentum).

A more precise formulation of these quantum rules in which Sommerfeld took an important part made it possible to specify all the possible orbits of an electron in the hydrogen atom in terms of several quantum numbers.

It was now possible to give a clearer answer to the old question whether Planck's postulate was a restriction on the energy of the radiation or of the radiative system. Bohr's theory clearly required that an atom or any closed system containing moving particles had discrete energy levels. If the Bohr–Sommerfeld quantum rules were applied to a system performing harmonic oscillations of frequency v, which figured in Planck's early considerations, they led to equidistant levels with an energy spacing of hv. At the same time it was essential for Bohr's theory of atomic spectra that the energy and the frequency of the light should be governed by Planck's relation. In other words, this relation applied to both the radiation and the radiative system. The universal character of the relation between energy and frequency is an important result which will later be seen to be essential for the development of the ideas of de Broglie.

The picture of the atom which emerged from these ideas was a strange and at first sight inconsistent combination of concepts taken from classical mechanics with new principles for which classical mechanics appeared to leave no room. The orbits of the electron were to be governed by the classical equations of motion, but of their many solutions only a very special set were regarded as possible, selected by rules foreign to the principles of classical mechanics. This also meant that one could not describe in detail the process of the emission or absorption of light during which the electron would have to change from one of the allowed quantum orbits to another, presumably passing on the way through orbits not permitted by quantum rules. Energy conservation could, of course, give us the total amount of radiation emitted or absorbed in such a condition, but without a detailed description of the process itself one could not obtain answers for such more detailed questions as the relative probability of transitions of an electron from one orbit to different final orbits, or the angular distribution and polarization of the light. Bohr was able to get a good deal of information on these questions from the 'correspondence principle', which relates some of the properties of the radiation with those expected from classical theory.

Classical physics must become correct for the larger orbits, when the quantum numbers are very large and the change from one to the next practically continuous. The requirement that in this limit the spectrum of the radiation should also approach that predicted classically leads to statements about the transition probabilities and about the associated radiation, at least in the limit of high quantum numbers, but the rule so obtained suggests extensions which may be applied also for all transitions.

It is not easy to summarize the argument involved in the correspondence principle in a few words, but the following simple example may indicate its spirit. Consider an harmonic oscillator, i.e. a charged particle attracted by a linear force towards a centre, its natural frequency of oscillation being v. Quantum theory then says that its energy must be, apart from an additive constant, equal to nhv, where n is an integer. If, in emitting light, it goes from a state n to a state n', it loses an energy of $(n-n')hv$, i.e. the frequency of the emitted light quantum is $(n-n')v$, i.e. some multiple of v. For large n and n' we approach the classical limit, in which we know that a system performing simple harmonic motion will radiate only with the fundamental frequency v and there are no higher harmonics. In this limit, therefore, there must be a selection rule, by which only the transition with $n' = n-1$ is allowed. This suggests the extension that in the light-quantum emission the harmonic oscillator can from any state go only into the next lower one, and this is indeed correct.

The success of Bohr's idea was so striking that, in spite of many unresolved difficulties, there could be no doubt that it contained a large part of the truth. Within a few years all studies of atomic problems were carried out in terms of Bohr's theory.

Amongst these successes there were, besides the explanation of the hydrogen spectrum, the theory of the X-ray spectra of the elements. The idea of discrete energy levels of atoms was confirmed in 1914 by the experiment of Franck and Hertz about the transfer of energy in collisions between electrons and atoms.

Above all, the general features of atomic structure and atomic spectra were becoming intelligible.

As a typical illustration of this, consider the alkali-metal atoms. The similarity of their spectra with the hydrogen spectrum suggested that one was dealing essentially with the orbits of a single electron to which definite quantum numbers could be assigned, and therefore that the inner electrons formed a stable system of orbits which underwent no substantial change in the usual optical transitions. These conclusions fitted in well with the chemical evidence which showed alkalis to be monovalent electropositive elements, and with the fact that the inert gases with one electron less than the alkalis were chemically inactive and optically inactive up to fairly high frequencies. Similar arguments could be extended to many other elements and led to an understanding of the periodic table in terms of the successive filling of the K, L, M, ... shells of the atom. However, at this stage it was not possible to account for the number of electrons that would fill a given shell, i.e. for the K-shell containing two electrons, the L-shell eight, and so on. It might have been thought that these numbers could be explained in terms of the mutual interaction of the electrons in each shell. A precise treatment of this problem would have required a calculation of the quantized orbits of a system of several electrons, a problem then of prohibitive mathematical difficulty, but it did not seem likely that such a simple set of occupation numbers could result from the solution of so complicated a problem.

Exclusion principle. Spin

In spite of the many impressive successes of the Bohr theory there remained some acute difficulties. One of these was connected with the fine structure of the spectral lines of the alkali metals. Each line was found to be a doublet, thus showing that there were twice as many quantum states in the atom as the theory predicted. The same difficulty exists already in the hydrogen atom, but the hydrogen fine structure was smaller and thus harder to observe, and also the picture was complicated by the special situation in hydrogen where orbits of the same principal quantum number, but different angular momentum, happen to have the same energy. It was at first believed that this doubling of the alkali states represented the existence of two alternative states of the inner shells or the 'core', but no explanation of this could be found.

More detailed information of the orbits was obtained from the Zeeman effect, i.e. the splitting of the spectral lines by a magnetic field. It is easy to understand that a magnetic field affects the orbits of an electron, and that the resulting precession depends on the orientation of the orbit in space relative to the magnetic field. The Zeeman effect thus concerned 'space quantization', i.e. the existence of orbits which were determined by the quantum conditions not only as regards their shape and velocity but also as regards their possible orientation relative to the field. The order of magnitude of the splitting also corresponded to the theoretical relation with the Larmor frequency, but the details were far more complicated than the theory would have predicted. The curious patterns obtained in the Zeeman effect and their variation with the field strength had been known for a long time, and Voigt had given an empirical formula which reproduced these effects in all details, but it seemed impossible to relate this formula to the Bohr theory.

A more direct verification of the space quantization was obtained from the work of Stern and Gerlach in 1922, which showed that beams of atoms were split into several components by an inhomogeneous magnetic field. Since the deflection of the beam in the field depends on the component of the magnetic moment of the atom in the field direction, it was thus established that this component could have a number of discrete values corresponding to the various orientations of the orbit permitted by the space quantization, but again the number of different states did not agree. (An experiment in 1927 showed that even a beam of hydrogen atoms was split in two components, thus showing that the hydrogen atom was capable of two states just like an alkali atom, and here there was no possibility of blaming this doubling on an inner electron core.) However, Pauli took the decisive step of relating this doubling of the orbit to the fact that each shell could hold twice as many electrons as there were orbits with distinct quantum numbers for that shell in the Bohr theory. Pauli recognized that this implied the existence of a further degree of freedom for the electron which gave it for any given orbital motion two possible states. On this basis the filling of the shells allowed just one electron for each possible quantum state specified by its orbital quantum numbers as well as the new degree of freedom. The principle that no two electrons could be in the same quantum state was called by Pauli the 'exclusion principle'. He realized that this was a new postulate to be added to the dynamical laws of the orbit and to the quantum conditions, and could in no way be derived as a consequence of the forces of interaction between the electrons or other dynamical effects.

Soon after this Goudsmit and Uhlenbeck suggested that the new degree of freedom might represent a spinning motion of the electron about its own centre. Since from the quantum rules an angular momentum of l units could have $2l+1$ orientations, a spin of one-half would give just two possibilities.

If it was further assumed that the ratio of magnetic moment to angular momentum for the spin was twice as large as for the orbital motion, so that the magnetic moment due to the spin angular momentum of half a unit was one full Bohr magneton, all the evidence about experiments of the Stern–Gerlach type as well as about the Zeeman effect were easily accounted for.

By the mid 1920's the Bohr–Sommerfeld quantum rules, supplemented by the postulate of the electron spin and by the exclusion principle, and used in conjunction with the correspondence principle, had led to an extensive description of very many of the facts about atoms and their spectra. Yet it was clear that there were still some essential elements missing from our complete understanding. In many ways the theory lacked consistency. For example, the wave properties of light and the resulting phenomena of diffraction and interference could still not be reconciled with the quantum hypothesis. There were difficulties of principle connected with space quantization. The orbits of an electron with a given angular momentum could have a discrete number of possible orientations, e.g. the 2p orbits of hydrogen or of an alkali could have three possible orientations, but it was not clear in general how to locate the appropriate directions in space. In the presence of a magnetic field it had to be assumed that the axis of such a 2p orbit, for example, could be parallel, at right angles, or antiparallel to the magnetic field, and this assumption corresponded to the evidence from the Zeeman effect or from atomic beams. But it was hard to imagine what would happen if the magnetic field was reduced to zero and replaced by a field in a different direction. As the field, however weak, changed from one direction to another the possible orbits would have to align themselves suddenly to the new directions. Similarly, the rules for space quantization provided no clear answer in the simultaneous presence of an electric and a magnetic field in different directions.

Furthermore, the quantum rules gave clear prescription for periodic orbits, but for the motion of a particle of sufficient energy to escape the attraction of the centre of force, it was not clear what quantum corrections to apply to the orbits calculated from classical mechanics. In other words, the theory was not capable of dealing with such problems as the emission of X-rays under electron bombardment ('bremsstrahlung') or the photoelectric effect.

These difficulties were resolved and the structure of the quantum theory of the atom completed by the development of quantum mechanics during the 1920s.

Quantum mechanics. Waves and particles. The uncertainty principle

The final formulation of quantum mechanics in its present form had its origin in three largely independent approaches. In 1924 de Broglie completed his doctoral dissertation in which he speculated about the analogy between light and matter. He started from the fact that there was a universal relation between frequencies and energy common to light and to particles in periodic orbits, at any rate evidently in the case of simple harmonic motion. Considerations of relativistic invariance led him to postulate that, as in the case of light, material particles also should have associated with them waves of a wavelength related to their momentum. He thus postulated that, for example, an electron of momentum p should belong to a wave of wavelength h/p. This statement was fully confirmed by the discovery, by Davisson and Germer, and by G. P. Thomson, of electron diffraction.

De Broglie noticed that, for example, in the case of circular motion such a wave could exist only if a whole number of wavelengths fitted the circumference of the orbit and that this led to a condition on the momentum of the particle in a circular orbit which corresponded exactly with one of Bohr's quantum rules. In this way the existence of discrete quantum states was seen to result from the same principle as the existence of characteristic frequencies in the vibration of a string or of a membrane.

Similar ideas were taken much further in the work of Schrödinger. His starting point was rather different from that of de Broglie, though he knew of de Broglie's work and no doubt appreciated the connection at an early stage.

Schrödinger started from the analogy between particle dynamics and optics, following in particular the point of view of Hamilton. He noticed that the differential equation of Hamilton and Jacoby for the orbits of a particle in a field of force could be regarded as an approximation to an equation of the form of a wave equation, which, for free particles, led to the same wave as the postulate of de Broglie. For the hydrogen atom, and for a number of other problems, the equation admitted solutions only for specific energies which agreed in the cases first studied exactly with those given by Bohr's theory. The equation also correctly admitted solutions for any energy when the corresponding classical orbit was not closed, i.e. above the energy required for escape of the particle. The Schrödinger equation also could immediately be applied to cases in which there were external electric or magnetic fields present and then gave uniquely the correct answers for the Stark or Zeeman effects. In the presence of an oscillating electromagnetic field the equation gave correctly the conditions of the particle which corresponded to the absorption and induced emission of light (though not at this stage the spontaneous emission).

Originally both de Broglie and Schrödinger were inclined to attribute physical reality to the new waves in the same sense in which the electromagnetic field of Maxwell is a real physical quantity. On this view the particle aspect of the problem would be something derived from the wave picture and only approximately valid. Indeed, there are some cases in which a localized wave train or 'wave packet' would remain closely localized for a considerable time and thus might approximately represent the behaviour of a single electron, but, in general, it would eventually disperse so that fractions of the electron would be found in different regions of space, contrary to the known physical behaviour of electrons.

It therefore became clear very soon that this literal interpretation of the wave field was not tenable.

The third line of development was initiated by the work of Heisenberg in 1925 and slightly preceded that of Schrödinger. Heisenberg started from the amplitude relating to atomic states as, for example, in radiative transitions from one state to another. Some information about such amplitudes is available from Bohr's correspondence principle, and Heisenberg attempted to obtain a set of rules which incorporated all this information, and led to a complete specification of these transition amplitudes including their time variation which would give the frequencies and therefore also the energy levels of the atom. His work and its further elaboration by Heisenberg, Born and Jordan led to such a complete set of rules, using the algebra of non-commuting quantities or matrices. It was found that, for bound states where one expected discrete states, these equations led to a unique prescription, and in those cases in which the equations could be solved exactly they gave answers in agreement with the Bohr—Sommerfeld conditions and also in all detail with the results of Schrödinger theory.

At first, the appearance of the very abstract mathematics of non-commuting quantities made this approach appear rather formal and remote from any intuitive physical understanding, and therefore less attractive than Schrödinger's version. Yet Schrödinger was able to prove the complete equivalence of the two approaches and to show that the matrix elements of Heisenberg were identical with certain integrals over the solutions of Schrödinger's equation. For example, the matrix elements of the coordinate of an electron appeared naturally in the expression for the centre of gravity of a wave expressed as a superposition of a set of stationary solutions of Schrödinger's equation. In other words, Schrödinger's and Heisenberg's form of the theory were just different mathematical formulations of the same physical laws.

In fact the knowledge of this equivalence was very helpful in getting to understand the physical meaning of either form. As regards the meaning of the wave function of Schrödinger, Born realized that the square of the wave function (i.e. the intensity of the wave) gave us the probability of finding the electron at a particular point. For example, in an electron-diffraction experiment it was impossible to specify the point at which a single electron would reach the photographic plate or other detector. All the theory provided was statistical information on how likely it was for the electron to be detected in a certain place. If the experiment was repeated with many electrons, as would be the case in a practical diffraction experiment, this would then lead to an intensity distribution corresponding to the predicted diffraction pattern. This idea was, of course, immediately applicable to the diffraction of light, and thus provided an answer to the old question of the relation between the wave and particle aspects of light, confirming the old idea of the 'guiding field' or 'ghost field'.

These considerations culminated in the formulation by Heisenberg, and later elaboration by Bohr, of the uncertainty principle.

This principle expresses the fact that any observation on a physical system must necessarily involve its interaction with a measuring apparatus, and this interaction may cause a change in the state of the system under observation. Whereas in classical physics, in which all such interactions proceed continuously, one can in principle make the interaction as weak as one wishes, and therefore reduce the disturbance of the system so as to make it unimportant, the occurrence of the quantum of action in atomic physics limits the extent to which the interference can be reduced.

There still arises no limit to the accuracy with which any one quantity can be determined, but the accurate knowledge of one quantity, for example the position of a particle, imposes a corresponding limit to the accuracy with which another variable, for example the momentum, can be known.

It is, therefore, never possible to determine the initial conditions for the motion with sufficient accuracy to predict the further time development exactly, and the resultant uncertainty in the orbit just corresponds to the difference between one quantum orbit and the next. One was forced to abandon the old deterministic view of physics in which the concept of the orbit of a particle appeared as something self-evident without any reference to the ways in which it could be made observable. It is, of course, not surprising that our intuition does not easily adapt itself to this situation, since it is developed from our everyday experience of space and time and motion for which quantum effects are completely negligible. In learning to live with these modified concepts physicists at least could benefit from the experience of similar adjustments that had been required by the idea of relativity. There, too, our intuitive concept of time had proved to be dependent on our everyday experience applicable to circumstances when all velocities are very small compared to that of light.

The new view of quantum physics foreshadowed in many of Bohr's ideas, but made more precise and consistent through quantum mechanics and the uncertainty principle, now showed the way of resolving the major conceptual difficulties. The impossibility of predicting the behaviour of a physical system in detail accounted for the occurrence of probabilities in the interpretation of wave mechanics. The old contradiction between the wave and particle views was also removed by the consideration that each of them contained a possible observation as the answer to a question, but that the two kinds of observation were mutually exclusive. For example, in a diffraction experiment in which a wave showed interference after passing through two slits in a screen, the particle concept suggests that the particle must have passed through one or the other of the slits, but by the uncertainty principle any observation designed to determine through which slit it had gone necessarily involved a sufficient change in the motion to destroy the interference.

To take another example, the conceptual difficulty about space quantization, which was mentioned previously, was removed by the fact that the uncertainty principle prevents us from knowing the plane of the particle orbit completely since the knowledge of one component of the angular momentum is not compatible with that of another. We may ask for the value of the component of angular momentum in a given direction (e.g. that of an applied field, weak or strong) and this would always be a whole number of units. Since we cannot specify more than one component at a time, the statement is still true for any axis we might choose, and the field direction comes into the argument only because the Zeeman effect in the field provides a means of measuring one particular component of angular momentum.

The occurrence of non-commuting quantities is now seen to be an appropriate mathematical tool fitting in with the uncertainty principle. Two non-commuting quantities cannot both be given numerical values and the restrictions resulting from this fact are precisely those imposed by the uncertainty principle.

At this stage, therefore, the mathematical formulation of the quantum theory of the atom and its physical interpretation were essentially complete, but there remained the task of developing techniques for applying the theory to particular situations and to generalize it to relativistic velocities and to the problem of subatomic physics, as we shall see in the second part of this article.†

† [The reference for the second part of this article, which is not reprinted here, is given on p. 15.]

2 O. R. Frisch

Take a Photon ...

O. R. Frisch, 'Take a photon ...', *Contemporary Physics*, vol. 7, 1965, no. 1, pp. 45–53.

When light falls on an ideal half-silvered mirror half of it goes through, the other half is reflected. This is readily understood if we consider light as waves: we get a transmitted and a reflected wave, each having half the original intensity. But light is also a stream of photons, each carrying the energy amount $h\nu$. The frequency ν of the waves is not changed by the mirror, so the individual photons in both the transmitted and the reflected beam must have the original energy $h\nu$. One concludes that the photons are not split but go into one beam or the other, at random.

If we recombine the two beams in an interferometer (Figure 1) we usually get interference fringes; because of a slight misalignment (often intentional) the phase relation of the two waves as they become superimposed varies from place to place and we get alternate reinforcement and cancellation of the two waves. But it is simpler to discuss an interferometer which is so adjusted that we get no fringes. For instance, if the two right-angle mirror pairs G–H and I–J are placed so that the intersection line of G and H is the exact mirror image (in the half-silvered mirror A) of the intersection line of I and J, then the light waves that pass through A on their return will cancel each other; all the light will go into beam 3 (the reflected beam), while beam 4 will have zero intensity.

But what happens to the *photons* in an interferometer? At first it was thought that interference occurred when two or more photons came together; but that was disproved when G. I. Taylor showed that interference fringes were formed just the same whether the light was strong or whether it was so weak that hardly ever two photons passed through the apparatus together.

It follows that single photons can exhibit interference, that 'a photon can interfere with itself'. It would seem that something does travel along both paths in the interferometer even when only one photon is admitted; but what is it?

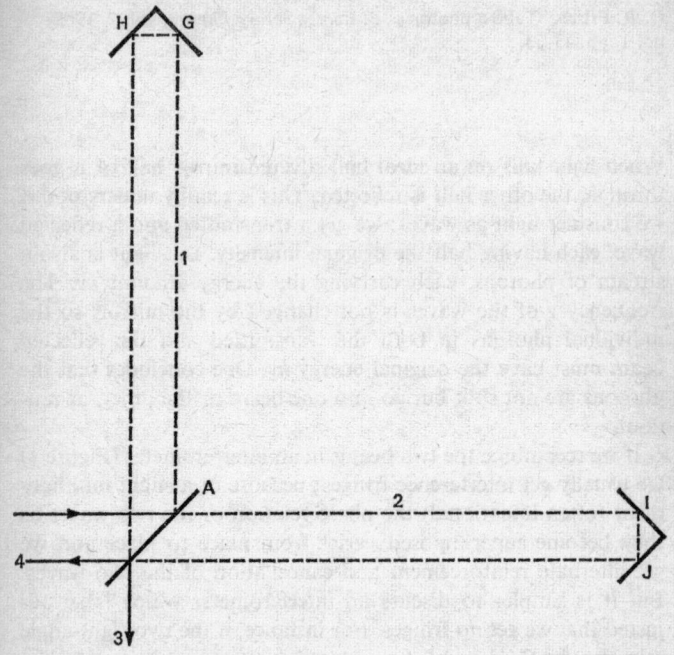

Figure 1 'Take a photon'

Such questions were discussed a good deal when photons were new, and similar questions arose out of the wave–particle duality of 'material' particles such as electrons. Some agreement has been reached on the way they should be answered, but the agreement is not unequivocal, and many of us are not sure what to tell our students. Indeed I am in two (or more) minds what to think, and for that reason I find it easiest to present the arguments in a dialogue between several characters. I have compelled them to be brief, but I'll try to elucidate what they say and sum up their findings at the end.

JIM Take a photon ...

TOM How?

JIM Well, take a weak light source and open a shutter long enough to let out one photon.

TOM But you may get two, or none!

BOB Is it single photons of visible light you want?

JIM Yes.

BOB Then I can help you; I have a generator for single photons of the sodium resonance line, $\lambda = 6 \times 10^{-5}$ cm. A beam of slow sodium atoms is crossed by a beam of yellow sodium light, which excites some of them. Those which emit a photon toward you are deflected on to a hot tungsten wire by their recoil, ionized and recorded by an electron multiplier. The output pulse tells us that a photon is on the way.

TOM Won't it be gone before we observe the pulse?

BOB There are lenses and mirrors to send the photon on a detour of 300 km, which gives you 1 ms notice; and there is a shutter which opens at the right time for an instant to let just that one photon pass through. The chance for another photon to arrive while the shutter is open is negligible.

JIM Fine. So we take a photon ...

TOM Can you make sure you have one?

BOB There is no need; my generator is quite reliable.

TOM Still, one ought to be able to make sure there is a photon.

JIM One could record it with a photomultiplier ...

TOM Not with any certainty!

BOB True; photocathodes have at best about 30 per cent efficiency. But with some semiconductors one can get close on 100 per cent. There is some noise, but with deep cooling ...

TOM All right; let us say we have a perfect photon detector. So we can make a photon, know when it will come, and verify that it has come. But in verifying we kill it!

JIM I know. Still, it seems I may at last say 'take a photon'. It behaves essentially like a particle: it starts from a point, it travels along a line, it ...

BILL Surely not! Light consists of waves; you can at best create a wave packet! And after travelling 300 km ...

BOB Let me give you the scale of my apparatus. My lens has a diameter a of one metre; over the distance $b = 300$ km the wave spreads by about $\lambda b/a$ which comes to 18 cm; the second lens is a little larger to allow for that, and it forms an image just as small as the original source, only a few wavelengths in size. The spread is ...

O. R. Frisch

BILL Essentially nil, I agree. But where does the photon pass through those large lenses of yours?

JIM I don't care; somewhere. Just let me take my photon from the focus of Bob's second lens. We can consider this as our photon source, and we know – we have 1 ms warning – when the photon is coming.

TOM All right, we'll let you take a photon. What will you do with it?

JIM I shall split it with a half-silvered mirror.

BILL But that doesn't split the photon; it is either reflected or transmitted, the chances being half and half.

JIM So the photon travels either in the direction 1 or 2?

BILL Sure. If you were to place photon detectors in both beams, either one or the other would record the photon.

JIM Good. Now please note that I have provided angle mirrors (Figure 1) which cause both beams to return and to be recombined.

BILL I see. You have built an interferometer similar to Michelson's. If your two distances are exactly alike then the reunited beam will go to the detector 3, not 4, if I've got the phase shifts right.

JIM Correct. From that I conclude that the photon has been split, that it is present in both beams 1 and 2. There is no element of chance. We need both parts of the photon to obtain the interference that causes it to be recorded at 3 and not at 4.

BILL But how do you account for the fact that of two photon detectors placed in 1 and 2, only one – at random – will record the photon?

JIM It must be the detectors that introduce the randomness.

BILL You mean that when one detector happens to report the photon, the other one is precluded from doing so?

JIM Yes.

BILL Even though the other half of the photon passes through it?

JIM No. Surely the other half no longer exists, it must have been destroyed when the photon was spotted in the first beam. Only one photon was produced by Bob's source, and a photon can only be absorbed and detected once.

TOM Isn't that the 'reduction of the wave packet' that the theoreticians talk about?

BOB I suppose so. But what does it mean? How can the observation of a photon in one place destroy the other half of the photon?

BILL There is no split photon; there is only a wave, which indeed is split.

BOB I must protest; my generator surely produces photons, one at a time. We agreed to that at the beginning.

ROY Let me try to remember what I was taught. The wave associated with one photon is not real; it is merely a mathematical tool which allows us to compute the probability that a photon will be observed at a given place. The split wave is just a description of our knowledge that a single photon has entered our interferometer and has met a half-silvered mirror. Once we know that the photon has been observed in one of the beams the probability that it should be found in the other becomes nil.

JIM Just as the chance for a horse to win a race becomes nil when another horse has won it?

ROY A bit like that.

TOM But if it is only a matter of probability that the photon is observed, might it not be missed by both detectors?

ROY Let us assume the detector in beam 1 is nearer the splitting mirror than the one in beam 2. Then if it records the photon it modifies the wave – which only represents our knowledge – so that 2 has nothing to detect. But if detector 1 remains silent at the critical time, then the wave in beam 2 gets strengthened so that the photon is sure to be recorded there.

TOM Which detector affects the wave if they are the same distance from the mirror, to within the length of the wave train? And anyhow, if the wave represents our knowledge it can only become modified by something that we come to know. What if we don't look at the first detector, but merely arranged for its signal to be recorded?

ROY Then there will be an even chance – just as if the first detector wasn't there – that the second detector will report the photon.

TOM Yes; but it will not report the photon in those cases where a later inspection of the first detector shows that it had, unknown to us at the time, recorded the photon. Does the present behaviour of the second detector then depend on the future state of our knowledge about the first one?

ROY No. We must interpret knowledge in a wider way. When one of the detectors records the photon, then the way it went is 'known' though you and I may not know it.

BOB This is getting ever more implausible: the knowledge stored in one box of electronics is said to affect a wave elsewhere – without signalling! – and so the behaviour of another box. Why not admit that the photon, on meeting the half-silvered mirror, takes a snap decision, at random, whether to go through or be deflected?

JIM Because then you can't account for the interference. If you were sure that half the photons travel along each path you could block up one of the paths and merely halve the intensity recorded by detector 3. But you know that if you block one path you destroy the interference: you then observe as many photons in detector 4 as in 3, whereas with both paths open all the photons arrive at 3.

BILL Wouldn't the wave account for the interference? There is both

O. R. Frisch 37

the wave and the photon. The wave gets split while the photon is either transmitted or reflected.

JIM But if we block one path with a detector and find the photon has gone that way, then you still have a wave travelling along the other path; a futile little wave without a photon! Unless you 'reduce it' and then we are back to where we were.

TOM Couldn't one spot the photon without absorbing it?

BOB Certainly. For instance, a transparent block of mass M, thickness a and refractive index n will be displaced forward by the amount $s = a(n-1)h/Mc\lambda$ when a photon of wavelength λ passes through it. That displacement is small, but . . .

BILL I know. However, such a block causes a phase shift of $2\pi(n-1)a/\lambda$ which will affect the interference and may destroy it.

BOB Can't we choose our block so that the phase shift is $N \cdot 2\pi$ where N is an integer? Then it won't affect the interference.

BILL Let me see. The displacement would come to $s = Nh/Mc$. If we measure to that accuracy we cannot know the momentum of our block to better than h/s (according to Heisenberg's uncertainty principle) or v to better than $h/sM = c/N$. If the block has the velocity v then the time the photon spends in the block is altered by the fraction $v/c = 1/N$, and so is the phase shift. So the phase is uncertain by 2π, and the interference is completely destroyed.

ROY Look, all this has been threshed out in Copenhagen, in the thirties: if you spot the photon you ruin the phase.

BOB I think I see a way around that. Let me suspend the half-silvered mirror so that we can measure its momentum perpendicular to its own plane . . .

ROY That has all been disposed of. If you measure the momentum of the mirror to within h/λ, the momentum of a single photon, its position is uncertain by λ, and that ruins the interference.

BOB No, wait, I have a new trick. I propose to suspend my mirror with a half-period equal to the time the photon takes to go out and back again along an arm of the interferometer. If it happened to deviate by $+d$ from its equilibrium when the photon first arrived it will deviate by $-d$ when the photon returns. Thus d has no effect on the phase, and the interference will not be ruined: it will happen just as if the mirror had been fixed in its equilibrium position!

ROY Ingenious. How will you measure the momentum transfer?

BOB Well, I just measure the momentum before the photon enters and again after it has been recorded at 3. The mirror will have been pushed one way if the photon was reflected at first and transmitted on the way back, and the other way if it followed the other path.

ROY But in the latter case the push comes half a period later than in the first; and the final outcome is the same whether you push a pendulum in one direction or half a period later in the opposite. So your measurement does not tell you which path the photon has taken!

BOB Ingenious. I fear you are right. So I must do one of my momentum measurements while the photon is in the interferometer, and that will spoil the interference.

TOM Still, you have designed a means of observing where the photon is, without doing anything to it; does that not prove that the photon really is in one beam, and not split?

JIM Einstein said something like that.

ROY You have not really observed where the photon is, merely by suspending the mirror; you must measure its momentum.

BOB But the momentum is in the mirror, and I can measure it or not, as I wish; surely that does nothing to the photon!

ROY Yes it does; it spoils the interference. All you have done is to share the clue about the position of the photon between the photon and the mirror so that it can be extracted from either of them. Your particular suspension ensures that this clue is automatically destroyed at the moment when the photon, by interference, gets directed to detector 3.

TOM If the mirror had not been suspended in that way but just left floating, then the momentum transfer would have measured the photon's position?

BOB Not necessarily. I think I could construct a mechanism by which you could make the mirror go to where the suspension would have taken it, if you so decide before the photon returns. All I need is ...

TOM We'll believe you. But once I have reflected a low-energy photon from the mirror so as to determine its velocity by Doppler shift, then the measurement is done?

BOB Not necessarily. I might construct a system of mirrors to send back the low-energy photon and return its momentum to the mirror.

TOM But when *is* the measurement done?

BOB I think any reversible process can be reversed. But I would regard myself as beaten if you have let the system interact irreversibly with say, a semiconductor, a photographic grain or a retina.

TOM That makes sense. After all, to measure is to create information; and information is a state – in a machine or an organism – which extends from a certain time into the future. Irreversibility is the very essence of information.

JIM But don't we sometimes obtain information without irreversible interaction? For instance, when the detector in beam 1 reports nothing we know that the photon is in beam 2.

BOB Yes; but the detector has to be there, in beam 1; the possibility of an irreversible interaction is essential.

JIM What if I place a piece of black paper into one beam? Then we have an irreversible interaction and we destroy the interference without getting any information in return.

BOB Not necessarily; you could measure the temperature of the paper before and after!

JIM But what if I don't?

BOB Oh well, information can get lost if you are careless.

JIM But, information aside, what does the photon do in my interferometer; does it get split, or doesn't it?

ROY You mustn't say 'information aside'. Quantum theory is about information. All it does is to tell you how to use available information to make the best possible predictions about future information.

JIM You mean, about what is going to happen?

BOB If you agree to use the word 'to happen' only for irreversible processes.

TOM Surely something happens – in the everyday sense – to the photon inside the interferometer; so quantum theory must be incomplete.

ROY I don't feel that. Quantum theory is logically consistent, and it allows you to make all the predictions that you can test. Photons and waves are models that allow you to use your imagination instead of using the full theory, but they cannot completely replace it.

BILL Couldn't one have a model that covers both photons and waves? Something more complex, perhaps multidimensional, of which our present concepts are merely flat projections?

ROY Plato's cave. Well, produce such a model, and we'll discuss it next time.

JIM But there are some worse difficulties which today we haven't even touched on. Take two photons . . .

Here we must break off the discussion and see what has been achieved. First we must admit that many of Bob's ambitious gadgets will never be built. That need not worry us; the use of thought experiments ('Gedanken-Experimente') in physical arguments has a long history and is generally accepted. Of course a thought experiment may be faulty if it contains an essential snag that has been overlooked by its designer. (A famous example was a thought experiment designed by Einstein and refuted by Bohr in 1930.) I have allowed Bob to explain his photon generator in some detail because photons are emitted at random, and some people have suggested that any thought experiment must be faulty if it assumes that we 'take a photon'. I think Bob has dispelled that idea. He cannot produce a photon at a specified time; but if Jim is willing to wait he will get one photon at a time and will know (a millisecond in advance) when it will arrive.

After that, the discussion turns to the question, what happens to the photon on striking the half-silvered mirror. Jim believes that it is split; Tom doubts it; Bill suggests that merely the wave is split. None of the attempted ways of visualizing what happens inside an interferometer appears satisfactory. The most common way is the one put forward by Roy, that the wave – which is split – determines the probability of the photon being intercepted in either beam 1 or 2. But then one has to assume that if the photon is found in, say, beam 1 the wave packet which represented the probability of finding it in beam 2 is thereby reduced to zero. This 'reduction of the wave packet' is not a physical process in the ordinary sense; it happens instantly, however far the two wave packets may be apart at the time when the photon is found in one of them.

Roy suggests that we should take the wave merely as a representation of what we know about the photon. Then the reduction process seems easier to understand; the wave packet becomes comparable to a list of betting odds which drop to zero as soon as a different contestant is known to have won the race. But it is not just a matter of our knowledge (say, that one detector has reported the photon) affecting our belief (concerning the chance of the other detector reporting it). We can place ideal photon detectors into both beams and record their signals; afterwards we can verify that every photon admitted was recorded by either one or the other of the two detectors. If we use the idea of a wave that is split we must assume collusion between the detectors (which is what the 'reduction of the wave packet' amounts to).

Yet we cannot simply assume – as Tom suggests – that each photon is either transmitted or reflected; if that were so we could not account for the interference, that is for the fact that all the photons emerge in beam 3 and not 4, provided both paths are kept open. If we want to visualize what causes the interference, we must think of a wave that is split by the half-silvered mirror and explores both paths.

Those two descriptions – a wave that is split and capable of interference, and a photon that is at random either transmitted or reflected – are called *complementary* aspects of the same thing, according to Niels Bohr. According to that view it is up to the physicist to choose that description which is appropriate to a given experimental situation. For the interferometer the wave picture is appropriate; if detectors are used to monitor the two beams (or even one of them) then the photon picture is appropriate. But there are intermediate situations (not discussed here) when neither picture is adequate, and then the mathematical theory must be wheeled into position.

The discussion then turned to the question whether the photon might not be spotted in one of the two beams without being absorbed. That is indeed possible (at least in the realm of thought experiments). For instance, a photon entering a block of glass changes its momentum from h/λ to $h/n\lambda$; the difference in momentum $(n-1)h/\lambda n$ is given to the glass block and causes it to move forward with the speed $(n-1)h/\lambda Mn$. After the time an/c the block stops again as the photon emerges from it; it has thus travelled the distance $(n-1)ah/cM\lambda$ (as Bob says).

But even though we have not absorbed the photon we have lost the chance of observing interference. Heisenberg's uncertainty principle tells us that the speed of the glass block will be uncertain by $\Delta v = \Delta p/M = h/Ms$ if we know its position to the accuracy s. That speed affects the time the photon spends in the block and hence the phase shift by just enough to make it uncertain whether the photon will go into beam 3 or 4. So once again, if we locate the photon we no longer observe the phenomenon – interference – which caused Jim to believe that the photon gets split.

At that point the inventive Bob comes up with a variant of the movable-mirror idea. His special suspension fails to allow him – as he first thought – to spot the photon and still to retain the interference. But it gives him liberty to choose: he can measure the momentum of the mirror (say by reflecting a low-energy – i.e. long-wave – photon from it and measuring the change, due to the Doppler effect, of its wavelength) before the photon has returned from one of the two angle mirrors, and that will tell him which way it went; or, if he does nothing, the mirror will swing in such a way that the interference is not disturbed. This is really a model of the EPR paradox (Einstein, Podolsky and Rosen†), namely the fact (demonstrated mathematically in that paper) that one can create situations in which one has the choice of measuring either the momentum or the position of a given particle 'without physically affecting that particle'. But that final clause (in inverted commas) doesn't really create a paradox once we have accepted that 'the same thing' – a photon that has met a half-silvered mirror – requires two totally different descriptions, depending on whether we have arranged to locate the photon or to observe interference. Again, there is nothing new in Bob's suspended mirror, merely a concrete example of something that is hard to think about. More severe forms of the EPR paradox exist and Jim wanted to talk about them, but I had to cut him off.

Perhaps the most important conclusion arises out of the insistence of Bob that 'any reversible process can be reversed', given enough ingenuity. The conclusion is that a measurement is not done until some irreversible process has taken place, such as an interaction with a grain in a photographic emulsion. Let me repeat what Tom says: 'To measure is to create information, which is a state – in a machine or an organism – which extends from a certain time into the future.' The intimate connection between information and irreversibility has indeed been stressed by L. Brillouin.

The main thing, as Roy points out, is that we must not ask what a quantum system does between observations. Quantum theory tells us how to use available information to make predictions about future information, and there are reasons to think (J. von Neumann) that they are the best possible predictions. I still feel a bit uneasy because I see no clear way of drawing the line between irreversible interactions which create (at least potentially) information and are to that extent 'real', and those interactions – like that between a photon and a half-silvered mirror – which don't create information and are 'unreal', demanding different descriptions in different circumstances.

As to Tom's last suggestion, of a possible model of which our present concepts – such as waves and photons – are merely shadows like those in Plato's cave, I have some sympathy with it; but no such model appears to be in sight.

†Einstein, A., Podolsky, B. and Rosen, N., *Physical Reviews*, vol. 47, 1935, p. 777.

3 P. Stubbs

The Introspective Photon

P. Stubbs, 'The introspective photon', *New Scientist*, vol. 39, 1968, no. 607, pp. 190–91.

The nature of light has exercised most of the very greatest minds throughout the history of science. Newton, who published his corpuscular theory in the *Opticks* of 1704, and Huygens, who described the rival wave theory in *Traité de la Lumière* in 1690, would surely both be delighted were they able to read the latest – and perhaps last – chapter of this great saga of dispute dating back at least to Democritus and Plato. For two physicists at the University of Rochester, Drs Robert Pfleegor and Leonard Mandel, by carrying out an extremely elegant experiment, now appear to have had the final say on the score of the dual particle–wave character of light. Their very fundamental observations vindicate a famous remark made some years ago by Dirac: '... each photon interferes only with itself', he said, 'Interference between different photons never occurs.'

On the face of it this statement may seem somewhat anomalous. The school textbooks traditionally tell us that interference between light beams is the simplest and most direct illustration of the wave aspect of light. We are taught of the classical experiment of Thomas Young in 1801 in which he passed light from a single source through a pair of thin slits and demonstrated that the two beams recombined to form a pattern of light and dark fringes on a screen beyond the slits. It looks simple. You throw two stones into a pond simultaneously and the ripples interfere, cancelling one another in some places and enhancing each other elsewhere – dark fringes and bright fringes. In 1865 Maxwell's electromagnetic theory, by showing how light waves came about, appeared to have clinched the wave hypothesis.

The arrival of the quantum theory in 1901, however, rocked the whole of nineteenth-century physics to its core. Planck revealed that light was emitted in discrete 'packets' or photons; Einstein that it was absorbed in the same fashion in the photoelectric effect; Compton that photons behaved like particles when they collided elastically with electrons.

The final conclusion was that light sometimes behaves as waves and sometimes as particles. Which aspect you see depends on what kind of experiment you do. Matter, since the quantum theory, is also regarded as having dual wave–particle characteristics. In electron diffraction, electrons interfere to produce a pattern analogous to that of Young's slit experiment. What is perhaps less well known is that Newton, for all his championing of the corpuscular theory of light, had considerable insight into the possibilities of such wave–particle combination theories which he never completely abandoned.

The new University of Rochester experiment has its roots in the Young experiment a century and a half earlier. We can imagine trains of waves interfering: but what happens if you cut down the light intensity so much that you are looking only at single photons? Do you see particle behaviour only, or do you still see wave behaviour as Dirac believed? Can a single photon interfere, in short, with itself?

Dr Mandel has done several experiments over the past six or seven years leading up to this crucial test. An important point about Young's experiment is that you cannot do it with two separate light sources instead of two slits – at least, not ordinary light sources. The reason is that the light is not 'coherent'. It is emitted from an ordinary thermal source (a lamp bulb, say) in irregular bursts. The waves from two such sources are far too erratic to produce regular interference patterns. Each burst from a single source, however, if split into two and recombined, contributes to a fixed pattern of fringes that depends only on the wavelength of the light and the geometry of the set-up.

When the laser was invented, its beautifully coherent waves following each other in highly ordered array, and their narrow bandwidth of frequencies, immediately suggested to optics researchers the possibility of repeating Young's experiment with two identical laser sources in place of the double slit. Five years ago Dr Mandel, then at Imperial College, London, tried it and it worked.

P. Stubbs

The success led him to ponder what was really happening at the fundamental level. Did the photons of one beam interfere with those from the second beam; or did each photon only interfere with itself? Chasing the answer has been a struggle to detect interference fringes from increasingly weaker pairs of laser sources.

In terms of photons, interference is a phenomenon that results in most of these 'light particles' arriving at bright fringes, and very few at the sites of dark fringes. If you want to look at the behaviour of single photons it is thus no good trying to photograph the fringes; you must count the photons arriving one by one at the places where you expect both the light and dark fringes to appear and then analyse the pattern of their arrival statistically.

In essence, the technique adopted by Pfleegor and Mandel is relatively simple. In practice, it is very tricky, partly because the coherence even of a laser beam is limited and confines the experimenter to a restricted 'exposure time' in which to make observations; and partly because the lasers get slightly out of tune with each other so that the positions of the interference fringes drift the whole time. In outline what the two physicists did was to direct the beams of two helium–neon gas lasers on to a single detector by means of mirrors so that they met at a small angle; *en route* each beam passed through an attenuator which sharply reduced its intensity until the number of photons in the beam was extremely small. Despite a rate of seven million photons per second falling on the detector, the attenuation was such that the time taken for a single photon to pass through the apparatus was only about one-fiftieth of the average time between successive photons. Statistically, therefore, one photon was nearly always received at the detector before the next was emitted by one or other of the lasers. As Pfleegor and Mandel put it: 'If interference fringes are formed under these conditions they cannot easily be described as an interference between two independent photons . . .'

For their detector they used a stack of thin glass plates placed edgewise to the beams – sliced up microscope–slide cover slips, each as thick as half the expected fringe width. Photons falling on the first, third, fifth, and so on, plates were registered by one photomultiplier; those falling on the second, fourth, sixth, and so on, on another. Clearly, if no interference occurred, both phototubes would count the same number of photons over a long enough period of time. If interference did occur, and the fringes stayed put, one tube would count nearly all the photons, the other very few. In practice, because the fringes drift about in position, you have to be a bit more cunning. What remains true, if interference occurs, is that when the number of photons recorded by one photomultiplier rises, that recorded by the other should fall. Put in mathematical parlance, there should be a *negative correlation* between the counts in the two channels if fringes are present. If they are not, there should be zero correlation.

Last year Pfleegor and Mandel reported preliminary results from their first experiment. By varying the angle between the two laser beams they could vary the expected fringe separation. In this way they found strong evidence for the presence of the negative correlation when the fringe separation corresponded, appropriately, to twice the thickness of the detector's glass plates. But the errors were rather large.

They have now repeated the experiment more precisely by automating the recording procedure. Punched-tape techniques have enabled them to study a much larger sample of interfering photons and the final answer is unequivocal: photons interfere essentially with themselves.

Just what does this statement mean? The physical presence of both lasers is necessary before interference will occur, even if only one of them is actually emitting a particular photon. A subsidiary experiment proved that the effect is one which 'is genuinely dependent on the strength of the optical field' and there is perhaps a clue contained in this phrase.

Electrons, in the equivalent of a Young's slit experiment, are known to behave in a similar way, each particle apparently passing through *both* slits and interfering with itself on the far side. One way of regarding particles of both matter and light is as the sum of an infinite number of continuous waves of all frequencies and amplitudes pervading the whole of space. Only at the point we call an electron or a photon do these waves add up to a finite value; elsewhere their sum is zero. According to this idea (due to de Broglie) although the particle or photon fades out ahead of and behind itself, it marks out what Professor Eric Rogers of Princeton University describes in his eminently readable *Physics for the Inquiring Mind* as 'guide ripples'. These are phase waves able to travel faster than light but not transmit energy.

Hence we might expect that each of our two lasers somehow distorts space in a manner which can be noticed by a single photon emitted from the other. The single interfering photon is thus, in one sense, partly a constituent of both beams.

Experiments like that of Pfleegor and Mandel go further than justifying important statements about the nature of energy and matter; they make it much harder for physicists to abandon the very basic concept of *fields* in space. Latterly, some theoreticians have moved back towards the idea of 'action at a distance' by which particles interact without the agency of any intervening electromagnetic or gravitational fields, or the strong and weak fields of high-energy particle physics.

4 A. Bairsto

Lasers

A. Bairsto, 'Lasers', *School Science Review*, vol. 46, 1965, no. 160, pp. 521–40.

The name 'laser' is formed from the initial letters of 'light amplification by the stimulated emission of radiation'.

A laser is a device which generates a beam of light. The light from a laser is very much more intense than light from other sources and, in addition, possesses the property of coherence which ordinary light beams do not possess. The angular spread of the laser beam is usually very much smaller than that of ordinary light beams.

In order to understand the working of a laser it is first necessary to consider the physics of the emission and absorption of light. Light is energy in the form of electromagnetic waves, and its emission involves the conversion into electromagnetic radiation of energy in some other form. Two of the characteristics of a light wave are its wavelength and its frequency. Light waves are very short, ranging from about 350 nm† for violet light to about 700 nm for red light. The frequency of a light wave is very high, ranging from about 4×10^{14} Hz for red light, to about 8×10^{14} Hz for violet ligth.

When the light from a discharge tube, such as a mercury vapour lamp, is examined spectroscopically, a line spectrum is observed, each line being formed by light of a single frequency. Thus the light emitted from an assembly of similar atoms consists of numerous discrete frequencies. In order to explain this, it is necessary to consider the conditions under which light is emitted.

Light originates in atomic systems, or in individual atoms. In an atom there are numerous possible orbits for the electrons. The energy possessed by an electron will depend upon which of these orbits it occupies. An electron may move from one orbit to

†[1 nm = 10^{-9} m].

another within the same atom and in so doing change its energy. In order to cause an electron to move into an orbit of greater energy it is necessary to supply it with energy from outside the atom. If an electron moves to an orbit of lower energy its surplus energy must be removed. The emission of a burst of electromagnetic radiation from the atom is the usual way of removing this surplus energy. Such a burst of radiation is of very short duration (for

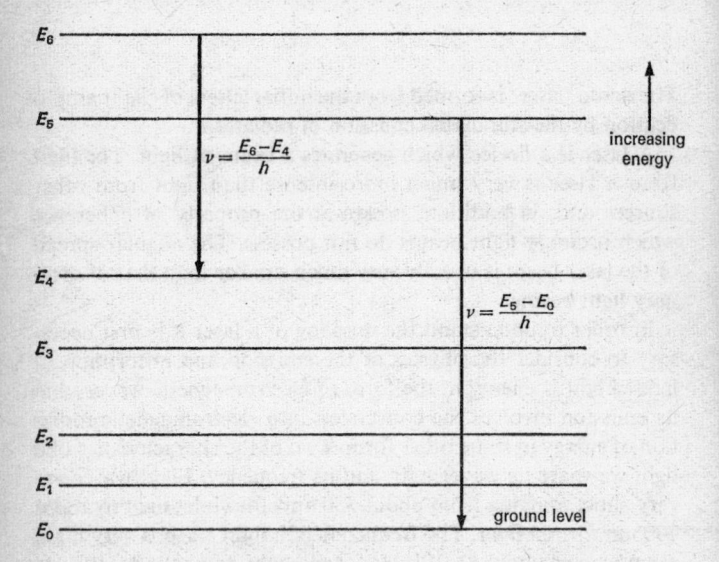

E_6 ─────────────────────────────

E_5 ─────────────────────────────   increasing energy

$\quad\quad\quad\quad v = \dfrac{E_6 - E_4}{h}$

E_4 ─────────────────────────────

$\quad\quad\quad\quad v = \dfrac{E_5 - E_0}{h}$

E_3 ─────────────────────────────

E_2 ─────────────────────────────

E_1 ─────────────────────────────

E_0 ─────────────────────────────   ground level

Figure 1 Energy-level diagram

example 5×10^{-8} s), contains about 10^6 waves, occupies a total length of wave train of about 50 cm and is known as a 'photon'.

When light is emitted its frequency is dependent only upon the quantity of energy which has been converted into radiation. The law $E = hv$ applies to the emission of radiation, where E is the energy converted into radiation, v is the frequency of the radiation and h is Planck's constant. The emission of a high-frequency light wave, such as blue light, requires more energy than the emission of a lower-frequency wave, such as red light.

The possible energies that an electron may have in an atom are known as 'energy levels'. These energy levels are numerous, but, for a particular type of atom, are definite and characteristic of that type of atom. In the consideration of electron transitions it is convenient to make use of energy-level diagrams (Figure 1). Each horizontal line represents one of the possible energy levels in the atom, the value of the energy increasing towards the top of the diagram. The lowest level is known as the ground level, or ground state, and the electron will normally occupy this level. Thermal energy is sometimes sufficient to raise electrons to levels slightly above the ground level, but much more energy than is available in this way is needed to raise electrons to the higher levels. One way in which an electron can obtain this energy is by the absorption of light.

Absorption of light

Light is emitted and propagated in complete photons, and it is usually absorbed in complete photons, or not absorbed at all. In order to remove an electron from an energy level inside an atom to a position completely outside the atom requires the

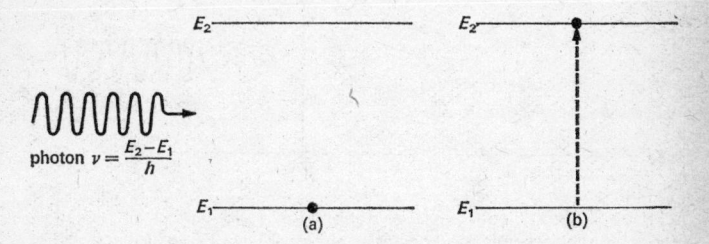

photon $\nu = \dfrac{E_2 - E_1}{h}$

(a)   (b)

Figure 2 Absorption of a photon. (a) shows a photon incident upon an atom in which an electron occupies the lower level E_1. (b) shows the electron lifted to the upper level E_2, the photon having been absorbed

absorption of a photon whose energy must, at least, be equal to the difference in the energy of the electron inside and outside the atom. If the photon energy is less than this value, the electron will not be disturbed and the photon will not be absorbed. If the photon energy exceeds this value, the whole photon will be absorbed, the electron will be removed from the atom, and the excess photon energy will be converted into the kinetic energy of the electron. If the energy difference is E, then radiation with a frequency equal to, or greater than, E/h is needed to eject an electron from the atom. However, in order to raise an electron from one definite energy level E_1, within the atom, to another definite level E_2, the photon must have exactly the correct amount of energy, $E_2 - E_1$. The radiation must, therefore, have exactly the correct frequency ν given by $\nu = (E_2 - E_1)/h$, and in lifting the electron between the two energy levels the photon will be completely absorbed (Figure 2). A photon of lower or higher frequency than this will not disturb the electron, and will not be absorbed.

Spontaneous emission of radiation

When an electron occupies a level above the ground level, the atom is said to be in an excited state, and is in an unstable condition. The electron, after a brief stay in the upper level, will spontaneously fall back to a lower level and will lose some of its

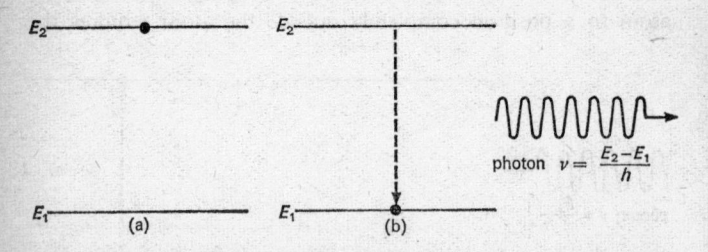

Figure 3 Spontaneous emission of a photon. (a) illustrates an excited atom in which an electron occupies an upper level E_2. (b) illustrates the emission of a photon when the electron spontaneously falls to the lower level E_1

energy. This energy will be emitted from the atom as a photon of radiation whose frequency will depend upon the energy difference between the two levels, *and will be exactly equal to the frequency of the photon which, when it is absorbed, would raise the electron between the same two levels* (Figure 3). It is this process of spontaneous emission which is mainly responsible for the emission of light from ordinary sources.

Within an atom, numerous electron transitions are possible, and each transition will involve a definite energy difference and hence a definite radiation frequency. Thus the light emitted from an assembly of similar atoms will consist of numerous discrete frequencies, or spectrum lines. Close examination of such a spectrum line, which is the combined result of similar electron transitions in numerous atoms, shows that there is a small frequency (and wavelength) spread (Figure 4). Each spectrum line

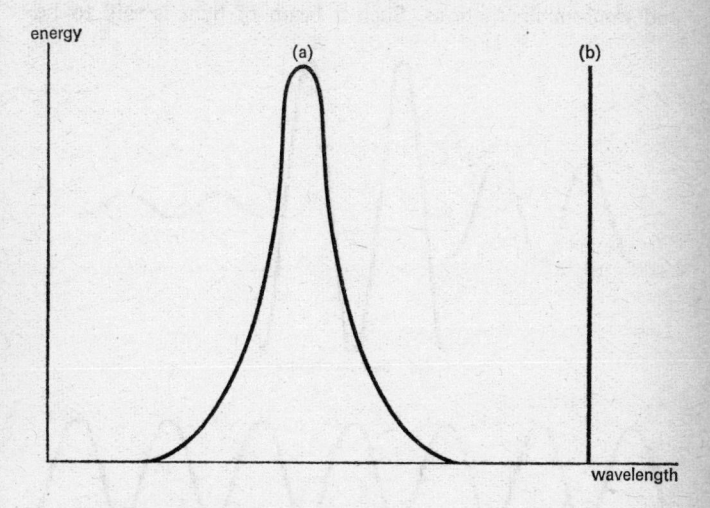

Figure 4 The width of spectrum lines. (a) shows the wavelength spread of a spectrum line compared with (b) which represents a truly monochromatic line. The spread of a spectrum line is usually less than 10^{-11} m

is really a very short continuous spectrum. Several factors contribute to the broadening of a spectrum line. Interaction between neighbouring atoms slightly modifies the energy levels themselves, and the motion of the emitting atoms causes an alteration of frequency by the Doppler effect. These are just two of several factors affecting spectral-line width. *This spread of frequency and wavelength plays an important part in the functioning of a laser.*

Incoherence of ordinary light

There are two aspects of incoherence: time incoherence and space incoherence.

Time incoherence

An ordinary light beam will have originated in many thousands of atoms and will consist of a large number of photons, or wave trains, superimposed. There will be no special relationship between these waves, and they will interfere and produce a resultant wave train whose amplitude and phase will suffer rapid and random fluctuations. Such a beam of light is said to be

(a) time incoherent

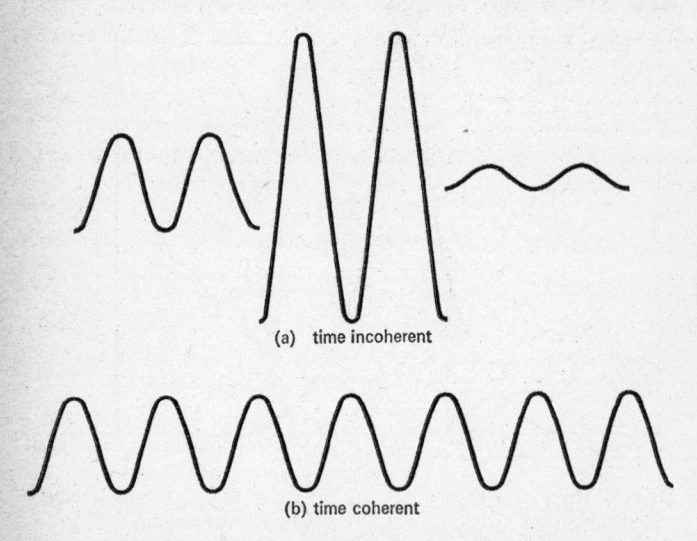

(b) time coherent

Figure 5 (a) shows a time-incoherent wave, as would be emitted from an ordinary light source, in which there are sudden jumps in phase and amplitude. (b) shows the time-coherent output of a laser

'time incoherent'. A beam which does not suffer such fluctuations is 'time coherent'. Figure 5 shows in a simplified form the essential difference between the time-incoherent output from an ordinary light source and the time-coherent output from a laser.

Space incoherence

Time incoherence is a property of a single beam of light, whereas space incoherence concerns the relationship between two separate beams of light. Two beams of light originating in different parts of a source will have been emitted by different groups of atoms.

Each beam will be time incoherent and will suffer random phase-changes, as a result of which the phase difference between the two beams will also suffer rapid and random changes. Two such beams are said to be space incoherent. Two beams are space coherent if they preserve a constant phase difference. This is possible, even when the two beams are individually time incoherent, as long as any phase change in one of the beams is accompanied by a simultaneous, equal phase change in the other beam. With ordinary light sources this is only possible if the two beams have been produced in the same part of the source.

Stimulated emission of radiation

An electron which has been lifted to an upper energy level will normally, after a short interval of time, spontaneously fall to a

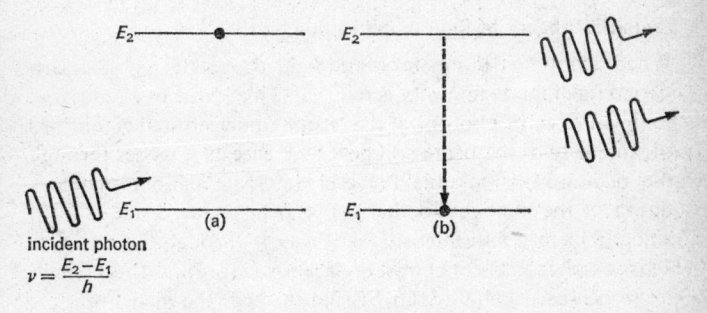

incident photon
$v = \frac{E_2 - E_1}{h}$

Figure 6 Stimulated emission of a photon. (a) shows a photon, of the transition frequency between the two levels E_1 and E_2, approaching an atom in the excited state. (b) shows the electron in the lower level, having been stimulated to fall by the passage of the photon. The photon emitted from the atom travels in the same direction as, and is in phase with, the stimulating photon

lower level with the emission of a photon of definite frequency v. If, whilst the electron is still in the upper level, a photon of exactly this frequency is incident upon the atom, the electron may be stimulated to fall to the lower level at the instant the external photon passes. The external photon is not absorbed or affected in any way, but passes on. In falling to the lower level the electron will lose energy, and this energy is emitted from the atom as a photon of frequency v, the same frequency as that of the external photon. Further, *this photon travels in the same direction as, and is in phase with, the external photon which stimulated its fall*. This is the phenomenon of stimulated emission of radiation, and the action of the laser is completely dependent upon it (Figure 6). To stimulate an electron to fall between two levels, E_2 and E_1, requires the incidence of a photon of exactly the correct frequency $(E_2-E_1)/h$. No other frequency will do. This is an effect predicted by Einstein, and will occur in company with spontaneous emission in ordinary light sources. For example, part of the light emitted from a sodium lamp will be spontaneous emission and part will be stimulated emission, the stimulating photons possibly having been emitted spontaneously.

A laser is designed so that stimulated emission overwhelmingly predominates over spontaneous emission.

Simplified theory of light amplification

When a beam of monochromatic light passes through a transparent medium its intensity is reduced. This is due to a reduction in the number of photons in the beam. Under normal conditions the intensity of the beam will never increase as it passes through the medium. It is possible, however, to create an abnormal condition of the medium so that a monochromatic beam of light actually increases in intensity as it travels through the medium. This is a phenomenon known as 'negative absorption'. The light beam has been amplified in passing through the medium.

In order to understand the conditions required for light amplification, consider a medium containing atoms in which there are two energy levels E_1 and E_2 ($E_2 > E_1$). The transition frequency between these levels is given by $v = (E_2-E_1)/h$. Suppose that in some of the atoms the electron occupies the lower level E_1, and in other atoms the upper level E_2. Let the number of atoms in each state be N_1 and N_2 respectively. Suppose that a beam of monochromatic light of frequency v (the transition frequency between the levels) is made to pass through the medium. Three different types of transition will take place:

(a) Some of the photons in the incident beam will be absorbed and cause electrons to be lifted from the lower level to the upper level. This process will cause a reduction in the intensity of the beam.

(b) Some of the photons in the incident beam will stimulate electrons to fall from the upper to the lower level, with the emission of another photon of the same frequency, the stimulating photon passing on unaffected. Since the stimulated photon is in the same direction as the stimulating photon, this process causes an increase in the intensity of the beam.

(c) Some electrons will spontaneously fall from the upper to the lower level with the emission of a photon. These photons are emitted randomly in all directions and, since only a small proportion will travel in the same direction as the incident beam, their contribution to the intensity of the beam can be neglected.

It is readily seen that, if the beam is to increase in intensity as it passes through the medium, the stimulated emission of photons must exceed the absorption of photons.

The number of upward transitions is proportional to the number N_1 of electrons in the lower level. The number of downward transitions is proportional to the number N_2 of electrons in the upper level. Einstein showed that the constants of proportionality for the two transitions are equal. Therefore, for stimulated emission to exceed absorption, N_2 must be greater than N_1: there must be more electrons in the upper level than in the lower level. Under normal conditions of thermal equilibrium, there are always more electrons in a lower level than in an upper level. The reversal of this condition is needed for light amplification. Such a condition is called an 'inverted population'.

Optical pumping

Let us now consider how an inverted population is achieved. In the ruby laser a method known as optical pumping is used. In order to bring about a condition of population inversion, it is necessary to increase the number of electrons in the upper level. If these electrons are removed from the lower level, the task of building an inverted population is made easier. A convenient way of doing this is to irradiate the material with monochromatic light, whose frequency is equal to the transition frequency between the two levels. This light is called the 'pumping radiation'. Photons will be absorbed and electrons will be lifted from the lower to the upper level. In order to achieve an inverted population, it is necessary to remove more than half of the electrons from the lower level. However, as the population in the upper level grows, the same radiation which is building up this population will begin to deplete it by the stimulated emission process. The more electrons that are established in the upper level, the greater is their rate of return to the lower level by the stimulating effect of the pumping radiation. It is, therefore, not possible to build up an inverted population by optical pumping between two levels.

Three-level system

The two-level system fails because the pumping radiation also stimulates the return of the electrons to the lower level. In the

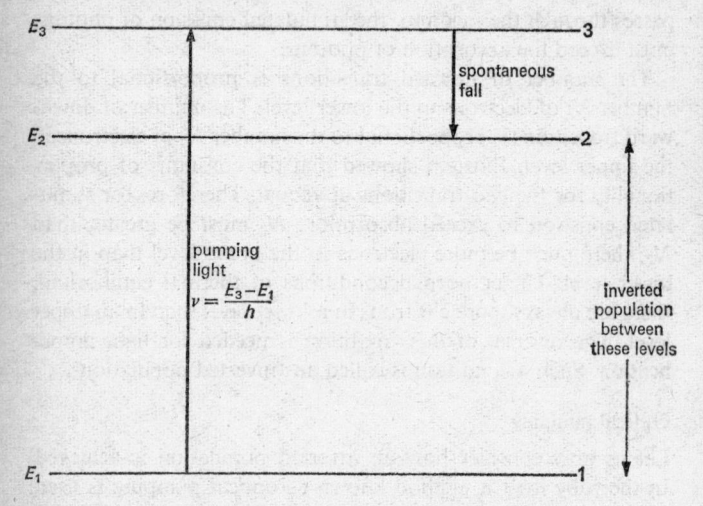

Figure 7 Energy levels in three-level laser. Radiation of frequency $(E_3-E_1)/h$ lifts electrons from level 1 to level 3. These electrons spontaneously fall to level 2, where they accumulate and build up an inverted population between levels 1 and 2

three-level system, radiation is used to pump electrons to a level above that in which they are required. The build-up of electrons is then brought about by their spontaneous fall into the required level from this higher level. Suppose that it is desired to build up an inverted population between levels 1 and 2 (Figure 7).

By irradiating the material with light of a suitable frequency $(E_3 - E_1)/h$, electrons are lifted from level 1 to level 3, from which they then spontaneously fall into level 2, where they accumulate until there are more electrons in this level than in level 1. It is clear that the three levels will have to be very carefully chosen. For example, the rate of spontaneous fall back to level 1 from level 3 must be much smaller than the rate of fall to level 2. Again, the average lifetime of electrons in level 2 must not be short compared with the lifetime of electrons in level 3. Otherwise electrons will spontaneously fall out of level 2 at a rate which does not allow a large population to be established. The advantage of the three-level system compared with the two-level system arises from the fact that the transition frequencies between levels 1 and 3, and between levels 2 and 1, are different. The pumping radiation is, therefore, not of the correct frequency to stimulate the removal of electrons from level 2.

Ruby laser

The first laser to work made use of a ruby crystal. Ruby is aluminium oxide containing a small proportion of chromium. It is the chromium atoms which are involved in the laser action. One of the features of the energy-level diagram of the chromium atoms (Figure 8) is a wide band of energy levels A. Consequently, light over a wide range of frequency can lift electrons out of the ground level into this energy band. This frequency range corresponds to light in the yellow–green part of the spectrum. The life of the electrons in this upper energy band is very short, and they fall spontaneously into a lower single energy level B. After a brief stay in this level, these electrons normally fall back spontaneously to the ground level with the emission of light of a single frequency. The wavelength of this light is 694·3 nm, and lies in the red region of the spectrum. This phenomenon is the ordinary fluorescence of ruby.

If a sufficiently intense beam of light strikes the ruby crystal, enough electrons will be removed from the ground level into the upper energy band, and from there into the upper single level, to produce an inverted population between this level and the ground level. When a beam of light of wavelength 694·3 nm passes through a ruby crystal in this condition its intensity will increase;

it will be amplified. It is not necessary, however, to pass an external beam of light through the crystal. By the normal spontaneous process, electrons will begin to fall back to the ground level with the emission of photons. These spontaneously emitted photons may stimulate the further emission of photons, and the

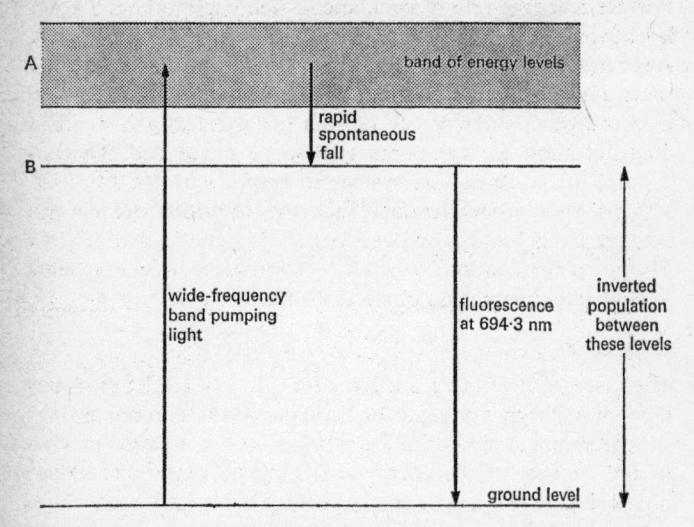

Figure 8 Energy levels in a ruby laser. This diagram shows some of the energy levels of the chromium atom. Light over a wide range of frequency in the yellow–green pumps electrons from ground level into the band A. A very rapid fall of these electrons to level B then occurs. Electrons returning to ground level from level B cause the emission of a line at 694·3 nm

intensity of the beam will grow. However, this beam would rapidly move out of the crystal and further growth of intensity would not occur. If the faces of the crystal are polished and silvered, the beam will be reflected to and fro many times within the crystal and, provided the amplification of the beam in each passage across the crystal exceeds the losses at the reflecting surface, the intensity of the beam will continue to increase. The process of stimulated emission will produce a chain reaction in which there is a very rapid return to the ground level of the electrons in the upper level. A very brief but intense pulse of light is emitted.

Since a photon emitted by the stimulated emission process is in phase with the stimulating photon, the cascade of photons produced by this chain reaction will all be in phase with each other, and the pulse of light will be time coherent.

The *energy* in this pulse can be no greater than the energy which would have been emitted spontaneously as normal fluorescence. The energy, however, is concentrated into a single pulse of very short duration, instead of being emitted randomly over a much longer period. The *intensity* of the beam is consequently very much greater.

Figure 9 shows the construction of one form of ruby laser. The

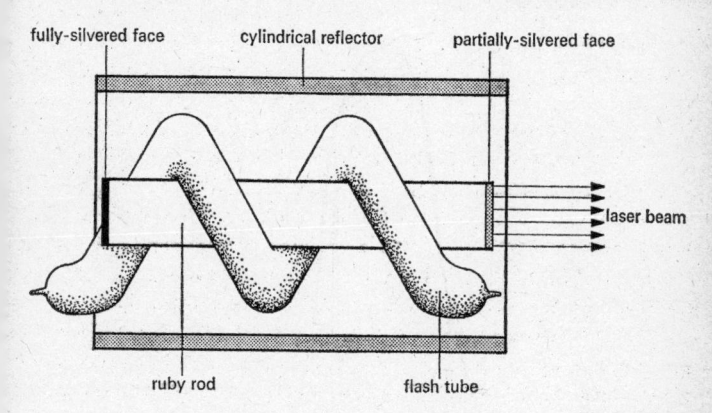

Figure 9 Essential components of a ruby laser

ruby is in the form of a cylinder several centimetres in length. The end faces are ground and polished flat and parallel. One end is fully silvered and the other partially silvered over its whole area, so that some light is reflected and some transmitted. A powerful electronic flash tube surrounds the ruby. After the flash is fired, the intense illumination causes the production of an inverted population in the ruby, and the processes already described lead to the generation of an intense pulse of coherent light.

Within the ruby the multiple reflections of the coherent light between the end faces result in a very large number of superimposed wave trains. These wave trains will neutralize each other by destructive interference, unless the distance between the

A. Bairsto

end faces of the ruby is exactly a whole number of half wavelengths. If this condition is satisfied, a very intense stationary wave will be produced in the ruby (Figure 10). The situation is analo-

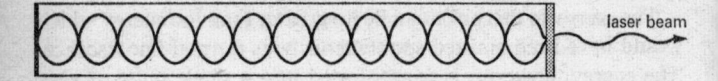

Figure 10 Stationary wave in the ruby crystal. Only that wavelength which satisfies the condition that the ruby length is a whole number of half wavelengths will be selected from the wavelength spread of the spectrum line, and be built up by stimulated emission to form an intense stationary wave

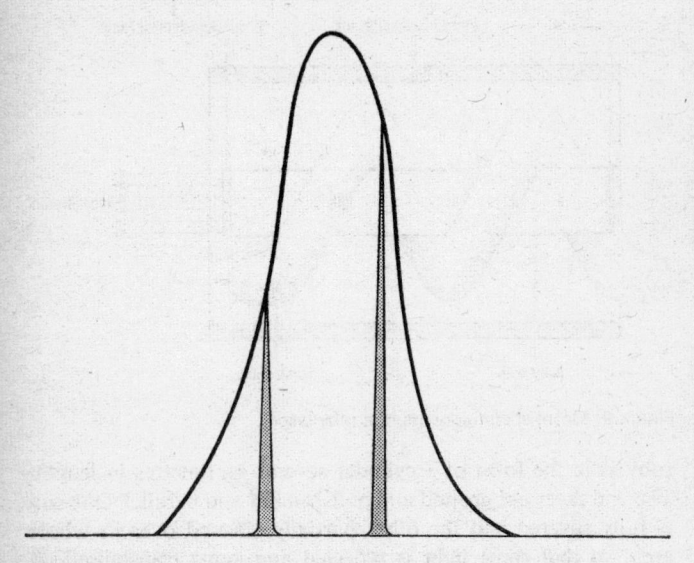

Figure 11 This diagram shows two wavelengths selected from the natural spectrum line width, and amplified by stimulated emission. The line width of the laser output is very much smaller than the natural line width

gous to the production of a stationary wave in a sonometer wire, or in an organ pipe. If the light emitted in the ruby were strictly monochromatic, the length of the ruby would have to be very accurately adjusted to satisfy this condition. In fact, the light emitted, although nominally of one wavelength, has a small wavelength spread (Figure 4). Only that wavelength which satisfies the condition that the ruby length is a whole number of half wavelengths will be selected from the wavelength spread of the spectrum line, and be built up by stimulated emission. Consequently the spectrum of laser light is a very much sharper line than that of spontaneously emitted light. There may even be more than one sharp line within the limits of the normal spontaneous line width (Figure 11).

The output of the laser is that portion of the light which passes through the partially silvered end of the ruby. If the exciting flash

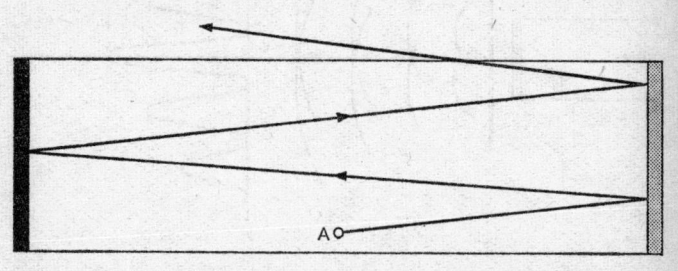

Figure 12 A photon emitted at A, and inclined only slightly to the cylinder axis will, after a few reflections, pass out of the side of the crystal with very little amplification

is not sufficiently intense to build up an inverted population, then amplification of the light in the ruby does not occur. However, the normal spontaneous return of the electrons to the ground level will take place, with the production of a relatively weak flash of incoherent ordinary fluorescent light in all directions.

The angular spread of the coherent laser pulse is very small. It is only those photons which are moving accurately parallel to the axis of the ruby cylinder which will be reflected back and forth sufficiently often to build up an intense beam. A photon inclined only slightly to the cylinder axis will, after only a few reflections, pass out of the side of the crystal with very little amplification (Figure 12). The output from the laser is, therefore, a very accurately parallel beam. If this beam is brought to a focus by a lens, an image area will be formed which is probably only limited in smallness by diffraction effects. A concentration of light energy is possible by this means far in excess of that obtainable in any other way.

The space coherence of the light emerging from the end face of the ruby may be demonstrated by performing Young's slits interference experiment using the laser as source. With ordinary light sources interference fringes can only be seen if the source is

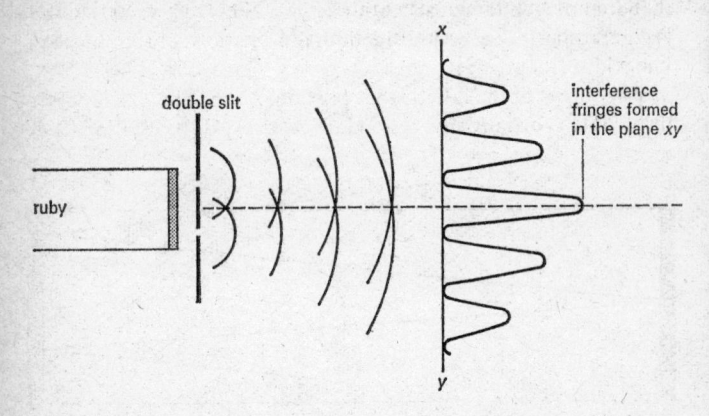

Figure 13 Young's slits experiment. Fringes can be observed even when the double slit is in contact with the face of the ruby, thus demonstrating the space coherence of the light from different points on the face of the ruby

very small. Fringes can be observed with the laser source even when the double slit is in contact with the face of the ruby, thus demonstrating the space coherence of the light from two different points of the source (Figure 13).

The production of a giant pulse. Q-switching

The output pulse from the ruby laser may last a length of time of the order of a few hundred microseconds. If the output is examined with a photocell and oscillograph, it is found to be of uneven intensity (Figure 14). In fact, the pulse consists of a large number of smaller pulses, or spikes. In order to explain this it is necessary to consider how an inverted population is built up in the ruby. The exciting flash of light lasts for a short time, of the order of a millisecond, and may be sufficiently intense to build up an inverted population in the initial stages of the flash. As soon as a sufficiently large inverted population has been produced, laser action will begin. This has the effect of depleting the upper level of electrons by stimulated emission more rapidly than they can be restored by the flash. The inverted population is thus destroyed and the laser action ceases. The flash is still active, however, and rebuilds the inverted population, whereupon the sequence repeats. A series of short pulses is produced until the intensity of the flash has fallen so that it can no longer rebuild the inverted population.

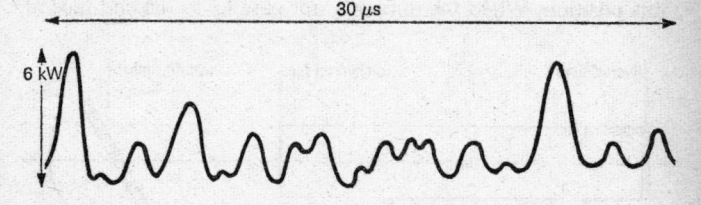

Figure 14 Pulsation in the output of ruby laser. If the output from the ruby laser is examined with photocell and oscillograph, it is found to be of uneven intensity and consists of a number of smaller pulses, or spikes

If a method can be found of preventing the depletion of the upper level until the exciting flash is over, a very large inverted population will be established. If the laser action is now allowed to commence, all the available energy will be concentrated into one giant pulse of extremely short duration.

The action of the ruby laser is dependent upon the repeated reflection of the beam between the end faces. The ruby will not work as a laser if these faces are not adequately reflecting, for the gain in intensity as the beam traverses the crystal is not sufficient to compensate for the losses through the end faces. If, therefore, the silvered coating is removed from one of the faces of the ruby, laser action will not take place, and there will be no premature destruction of the inverted population. If the reflecting coating could be replaced after the large inverted population had been established, then laser action would commence, and the whole of the energy stored in the ruby would be emitted in one giant coherent pulse.

A. Bairsto

There are several practical ways of carrying out this procedure, which is known as Q-switching, and a simplified diagram of one arrangement is shown in Figure 15. One end of the ruby carries a normal full coating of silver, and the other end is unsilvered. Opposite the unsilvered end a mirror is placed which is rotated at high speed about an axis perpendicular to the diagram. When the mirror is exactly parallel to the end face of the ruby, it will perform the same function as the normal silver coating on that face, and cause the light beam to be repeatedly reflected through the crystal. Thus laser action will take place with the mirror in this position. When the mirror is not parallel to the end face of

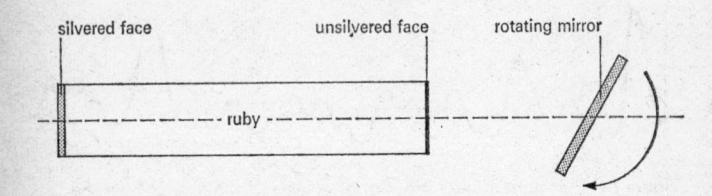

Figure 15 Production of a giant pulse. Q-switching technique

the ruby, the reflected light will not make sufficient traverses of the crystal, or will not even be reflected back through the crystal, and laser action will not take place.

The firing of the exciting flash is synchronized with the rotation of the mirror. The timing is such that the flash is fired shortly before the mirror becomes parallel to the end face of the ruby. The inverted population will have reached a very high value by the time the mirror is parallel to the end face. At this instant the laser action begins, and a single giant pulse of light is produced.

Early experiments gave giant pulses with a peak power of about six hundred kilowatts, lasting for about $0.2 \mu s$ (Figure 16). This compares with a peak power of about six kilowatts lasting for a few hundred microseconds when the ruby laser is operated in the ordinary way. These enormous powers are, of course, the result of the extremely short duration of the pulses. The total energy carried by the pulse is not large. In the example quoted above the energy of the giant pulse would be about one joule. The total energy output from the laser cannot exceed the energy of the exciting flash, and will be very much less, since much of the energy of the flash is converted directly into heat. Later experimenters have considerably increased the peak power of the giant pulses. A figure of ten megawatts is quoted in one instance, and this may well have been exceeded since.

The ruby laser will only work intermittently because the intense pumping radiation can only be produced in intermittent flashes and, because much heat is generated in the ruby, time must be allowed for the ruby to cool down.

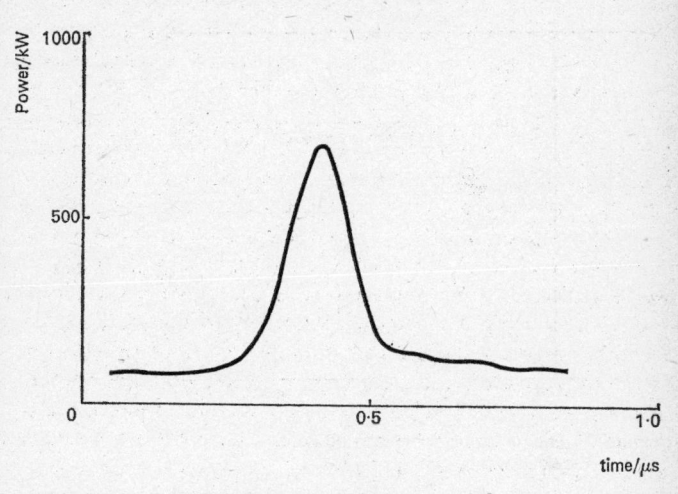

Figure 16 A giant pulse from a Q-switched ruby laser

Gas lasers

Laser action is possible in a gas or gas mixture. The production of an inverted population in a gas presents features which are not present in the case of a solid. The optical pumping of ruby is dependent upon the existence of a broad band of energy levels, so that light over a wide range of wavelength is absorbed. In gases such broad absorption bands do not exist. A gas absorbs only discrete wavelengths. It is possible to pump a gas optically using monochromatic radiation. However, a more convenient method is possible. The atoms in a gas may be raised to excited states by passing an electrical discharge through the gas. The discharge, however, is not selective and atoms will be raised into all possible excited states. A dynamic equilibrium will be established in which the number of atoms in each excited state remains constant. Some atoms will leave the excited state, others will enter it, and if these processes neutralize, the number of atoms in the state remains constant. It is possible that this process will result in a greater population in a high level than in a lower level to which

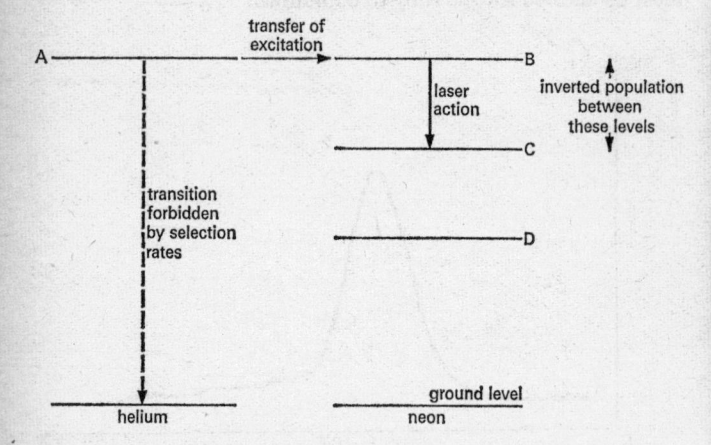

Figure 17 Energy levels in helium and neon

a transition can be made. An inverted population is thus produced. However, early workers did not succeed in attaining laser action this way.

The first gas laser to work used a mixture of helium and neon. Neon is the laser material. A simplified energy-level diagram for helium and neon is shown in Figure 17. The lowest energy level of helium above the ground level is A. Helium atoms accumulate in this level because direct transitions back to the ground level are forbidden by the quantum selection rules. An inverted population could be produced between these levels, but, since downward transitions are not possible, this would not lead to laser action.

The energy-level diagram for neon shows a level B approximately equal in energy to that of level A of helium. A process of transfer of energy can take place when a helium atom in the excited state A collides with a neon atom in the ground state. The helium atom returns to the ground state, and the neon atom is raised to the excited state B. The normal processes of the discharge will have produced a distribution of neon atoms amongst the various possible excited states. The transfer of excitation from the helium atoms raises the number of neon atoms in state B to an abnormally high value, and it is possible that an inverted population will be established between level B and level C.

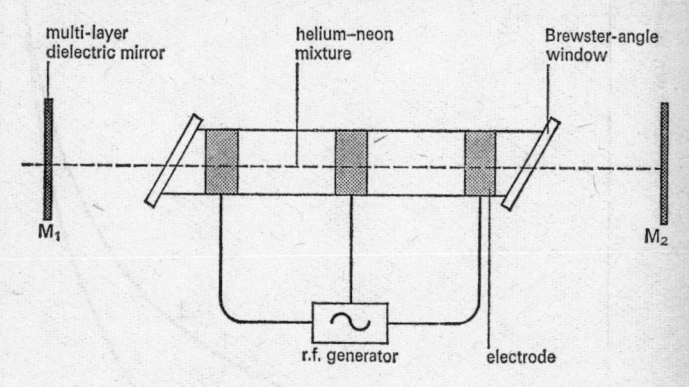

Figure 18 Diagram of a gas laser

Once this inverted population is produced, laser action can take place continuously, since the electrical discharge can be maintained. As fast as electrons are removed from the upper level B by stimulated emission, they are replaced by the transfer of excitation from helium atoms. It is important in maintaining the laser action that electrons which have fallen from level B to level C should not accumulate there, otherwise the inverted population would be destroyed. In fact, there exists a lower level D to which the electrons from level C rapidly fall.

A simplified diagram of a gas laser is shown in Figure 18. The light beam is reflected backwards and forwards through the gas between mirrors M_1 and M_2. In the first gas lasers the mirrors were placed inside the discharge tube. In more recent lasers the mirrors are placed outside the discharge tube, as shown in the diagram. The greater atomic separation in a gas, compared with that in a ruby, necessitates a longer light path in the gas in order that the amplification in each transit should be more than sufficient to compensate for the losses at the mirrors. The usual length of a gas laser is about one metre. The mirrors are multilayer dielectric mirrors with a reflection coefficient of about 99 per cent for the particular wavelength emitted by the laser. The output from the laser is the 1 per cent of the beam which passes through the mirrors. The excess of gain over loss in a

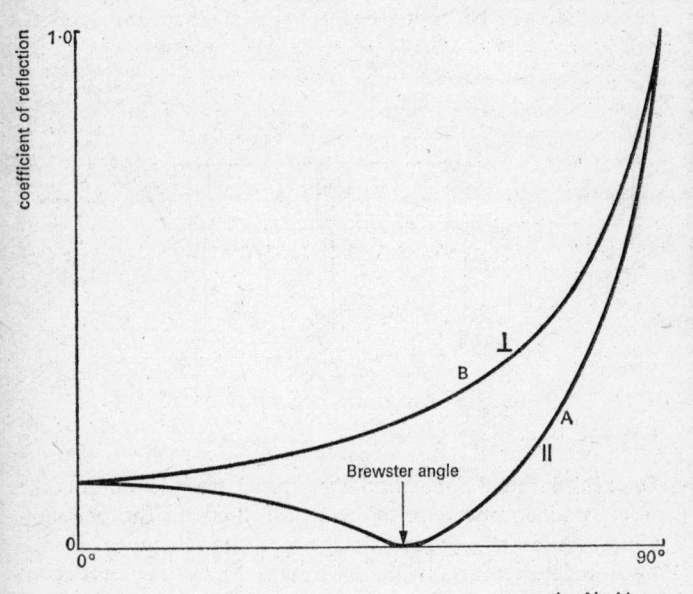

Figure 19 Diagram showing the variation of reflection with angle of incidence and the state of polarization of the light

single transit between the mirrors is so small that a slight increase in the losses will stop the laser action completely. For example, a puff of cigarette smoke in front of one of the mirrors causes sufficient extra absorption to stop the laser.

It will be noticed that the end windows of the discharge tube are not perpendicular to the axis of the tube. This arrangement reduces the loss of light by reflection at the windows. The proportion of a beam of light reflected from a transparent surface varies with its angle of incidence and with its state of polarization. Figure 19 shows how the proportion of the light reflected varies with angle of incidence from 0° to 90°. Curves A and B refer to the direction of the electric vector parallel and perpendicular to the plane of incidence, respectively. Curve A shows that when the electric vector is parallel to the plane of incidence the proportion of the light reflected is zero at one particular angle of incidence. At this angle the light will be transmitted without loss. This is the Brewster angle and is the angle at which the discharge tube windows are set. Any light in which the electric vector is parallel to the plane of incidence will be transmitted through these windows without loss. Light beams, in which the electric vector is perpendicular to the plane of incidence, will suffer some loss of intensity by partial reflection, and this loss is sufficient to prevent the establishment of laser action for this light. The light output from the gas laser is therefore plane polarized.

Applications

There are three properties of a laser beam which distinguish it from light beams from other sources:

1. It is time and space coherent.
2. It has, usually, a very small angular spread.
3. It has a very high intensity.

Proposed and actual applications of the laser make use of these features.

The coherence of the output has led to the suggestion that it may be used in communications. The output from a gas laser resembles, but at a very much higher frequency, a radio carrier wave. If the laser beam could be modulated, as a radio wave is modulated, it should be possible to transmit information along the beam. The number of separate channels which can be transmitted on the same carrier wave increases with increasing carrier frequency. The advantage of the laser beam is that its frequency is about a million times higher than a television carrier wave and it is, therefore, capable of carrying much more information. It has been estimated that a single laser beam could carry all the existing channels of information between the east and west coasts of the USA. The disadvantage of the laser beam for this purpose is that it cannot be transmitted long distances through the atmosphere because of absorption by mist and rain. Experiments have been conducted into the passage of laser beams along the inside of silvered pipes in order to overcome this difficulty. Before the method could become a practical possibility, means would have to be found to modulate and demodulate the laser beam.

The very small angular spread of the beam could be utilized in light signalling. The laser would be specially useful in space communication, where there is little absorption. The lack of absorption, combined with the small angular spread of the beam, would result in very little change in intensity of the beam even over very long distances.

Several uses have been found for the high-intensity output of the pulsed ruby laser. When the light from such a laser is focused by a lens, extremely high concentrations of energy can be produced. If an absorbing object is placed at the focus of the beam, the energy of the beam will be converted into heat. Although the heat energy is quite small (with the one-joule giant pulse quoted earlier the heat produced would be about a quarter of a calorie), it is concentrated into an exceedingly small volume and it causes a very large localized rise in temperature. This may be sufficient to cause local melting of the material. One use envisaged for this type of laser is micro-welding, and the drilling of small holes. Very small holes have been drilled through diamonds and through razor blades by this method.

The reattachment of detached retinas is a medical example of this technique. The laser beam is directed into the eye of the patient through the eye pupil. No harm is done to the eye lens or cornea because they are transparent, and the beam is not sufficiently concentrated. The eye focuses the beam on to the retina where it becomes very concentrated and where it is absorbed. The localized heating is effective in 'welding' the retina back into position. Laser beams have also been used for the very localized destruction of unwanted tissue in the human body.

The combination of high intensity and small angular spread of the beam has been used to produce a range-finder working on the radar principle. The laser beam is directed at the distant object. The beam is so intense that the very small proportion of the light which is scattered back to the instrument from the object is detectable. The range of the object is determined from the time interval between sending out the pulse and receiving the 'echo'.

There are many applications in research which cannot be classified. The investigation of optical effects, which, with ordinary light beams, are very faint and difficult to observe, is an obvious application.

It has been suggested that lasers might be used in teaching to demonstrate more easily than is possible with ordinary light sources such phenomena as diffraction and interference.

Much work is being done in the field of lasers. However, at the present time it does appear that lasers have not yet found the many applications that were envisaged when the first lasers were made to work.

Part Two
Nuclear Physics and Fundamental Particles

The papers in this part show that present knowledge has taken us a long way from the first theories of nuclear structure when the only fundamental particles were the proton and the electron, and later the neutron and the neutrino. As higher- and higher-energy particles have become available from larger and larger accelerators the picture has become increasingly more complex. This is a field of investigation which is very much alive at the moment and fresh discoveries are frequently being reported. It is therefore essential in reading any literature on the subject to note when it was written and to realize that it may be more or less out of date. I have tried not to include ideas which are too speculative at the moment but nevertheless the reader should bear in mind that some of what he reads here may not be up to date when he reads it.

The first paper by Blin-Stoyle is a summary of the properties of the nucleus and of two of the models used to describe its behaviour. To help in following the arguments given in section 5 on the shell model, some further extensions of the quantum-theory ideas given in Part One are needed.

We saw that in an atom the electrons have only certain discrete energy values, referred to as energy levels. These were first predicted on the basis of evidence from atomic spectra and later confirmed by the application of wave mechanics to the structure of atoms. In order to describe the particular level in which an electron is, certain quantum numbers are used. The first of these, called the principal quantum number and denoted by n, was first used in Bohr's theory to describe the energy of the electron. It can have the integral values 1, 2, 3,..., each number indicating successively higher energy levels. Electrons with the same value of n are said to occupy the same shell.

It is also necessary to specify the angular momentum of the electron in its orbital, and this is also subject to quantum conditions. For this a quantum number l is introduced. l can have the values $0, 1,..., n-1$. Following the notation used in spectroscopy the value of l is often indicated by the letters s, p, d, f, g,.... Thus, for example, describing an electron as 2p means $n = 2$, $l = 1$. Electrons with the same n and different values of l have slightly different energies.

The early ideas of space quantization (i.e. that the angular momentum of the electron can have only certain orientations with respect to the direction of a magnetic field) lead to the introduction of a third quantum number m known as the magnetic quantum number, and taking the $2l+1$ values $-l,..., 0,..., +l$. This quantum number also arises from the wave-mechanical treatment. For example the d-electron, which has $l = 2$ can have $m = -2, -1, 0, +1, +2$.

The discovery of electron spin resulted in the introduction of yet another quantum number s, which can have the values $+\frac{1}{2}$ or $-\frac{1}{2}$ depending on whether the spin is parallel or anti-parallel to the orbital momentum.

It is often convenient to combine the orbital and spin angular momenta quantum numbers (l and s) to produce the total angular momentum quantum number, denoted by j; j can have the values $l\pm\frac{1}{2}$ but it cannot be negative. Thus for an s-electron ($l = 0$), $j = \frac{1}{2}$ only; but for a p-electron ($l = 1$), j can have the values $\frac{1}{2}$ or $\frac{3}{2}$. The notation used to describe the quantum state of an electron is then, for example, $3p_{\frac{1}{2}}$, meaning $n = 3$, $l = 1$, $j = \frac{1}{2}$. The magnetic quantum number (now m_j) can then have any of the $2j+1$ values from $-j$ to $+j$.

In section 5 of his paper Blin-Stoyle examines whether these ideas can also be applied to describe the particles in the atomic nucleus as well as to the orbital electrons.

The two papers by Stannard review some aspects of high-energy nuclear physics including some indication of the large number of so-called fundamental particles which have been discovered. There is some speculation that the present 'fundamental particles' are in fact combinations of other even more fundamental particles – the quarks – but there is not at the time of writing (1970) any firm evidence for this.

The neutrino has proved to be the most difficult of the particles to observe because of its very rare interaction with matter. The paper by Ramm explains how experiments with high-energy neutrinos have been made possible and outlines some of the results obtained from this research. More progress can be expected in this direction in the near future.

5 R. J. Blin-Stoyle

The Structure of the Atomic Nucleus

R. J. Blin-Stoyle, 'The structure of the atomic nucleus', *Contemporary Physics*, vol. 1, 1959, no. 1, pp. 17–34.

1 Introduction

It is now nearly fifty years since Rutherford firmly established from the way in which charged particles are scattered by matter that an atom consists of a massive central nucleus surrounded by a cloud of electrons, the total electric charge of the electrons being equal and opposite to that of the nucleus so that the atom as a whole is electrically neutral. He further concluded that whereas the electron cloud has a spatial extension of the order 10^{-8} cm, the nucleus is a very small structure confined to a region of dimensions a few times 10^{-13} cm. Thus in most physical and chemical processes the nucleus can be regarded as a point charge; only the electrons take part in the interactions.

Since Rutherford's work a great deal of effort has been expended in investigating the nature and structure of the atomic nucleus. In the space of an article of this size, however, it is quite impossible to give any account of the ingenious and painstaking experiments that have been performed. Suffice it to say that the apparatus has varied from the crudest 'string and sealing-wax' arrangements to the gigantic high-energy machines such as the bevatron and cosmotron. Some experiments have been extremely inaccurate, others of the utmost accuracy, but all have contributed to a now well-established body of knowledge about the nucleus.

2 Properties of the nucleus

2.1 *Mass*

Nuclear masses are roughly integer multiples of the mass of the proton (the hydrogen nucleus) the integer being denoted by A and referred to as the mass number. Nuclei with values of A

from 1 up to 255 have been observed although a great many of them do not occur naturally.

2.2 Electric charge

The charge of a nucleus is always an integer multiple (Z) of the fundamental unit $e = 1\cdot60 \times 10^{-19}$ C and is written Ze where Z (the atomic number) is also equal to the number of electrons in the surrounding electronic cortège. Now it is frequently found that several nuclei having different values of A will have the same Z. The corresponding atoms have essentially the same chemical properties since these are determined by the electronic structure which will, of course, be the same in each case. Such nuclei are called *isotopes* and may have very different nuclear properties.

2.3 Size and shape

Most nuclei are nearly spherical in shape with a radius R given approximately by the formula

$$R = r_0 A^{\frac{1}{3}},$$

with $r_0 \simeq 1\cdot2 \times 10^{-13}$ cm. The density distribution $\rho(r)$ is not

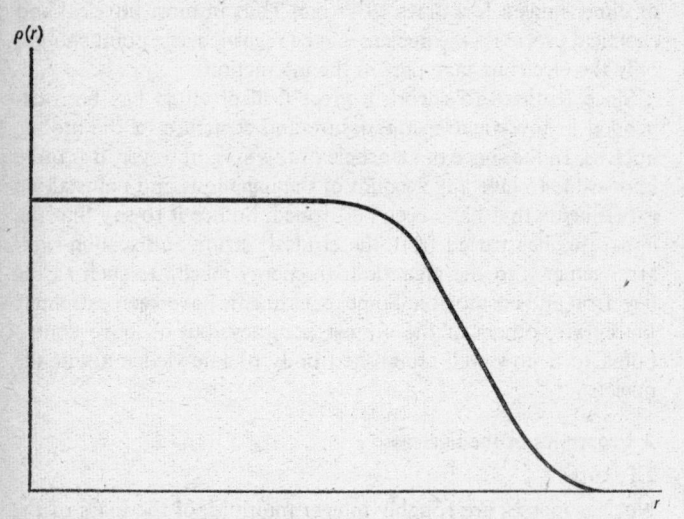

Figure 1 The density distribution $\rho(r)$ of a typical nucleus plotted as a function of r (arbitrary units)

constant however, and varies in the radial direction in the manner shown in Figure 1. This means that there is no sharply defined surface to the nucleus and that the radius R must be defined in some average sense. A suitable definition is that R is the root mean square radius, given mathematically by

$$R^2 = \frac{\int_0^\infty \rho(r) r^4 \, dr}{\int_0^\infty \rho(r) r^2 \, dr}.$$

2.4 Spin

In classical mechanics the idea of a spinning body is perfectly familiar and there is no restriction on the angular momentum (or spin) that it may possess. However, when we are dealing with a structure as small as the nucleus, then the angular momentum will also be very small and quantum-mechanical effects become important. Quantum mechanics restricts the angular momentum of a system to be an integer or half-integer multiple of a natural unit of angular momentum $h/2\pi$ (often denoted by $\hbar$), where h is Planck's quantum constant having the numerical value $h = 6.626 \times 10^{-34}$ J s.

Experimentally it is found that a great many nuclei have an intrinsic spin $I\hbar$ with I, the quantum number, taking all integer and half-integer values up to $\frac{9}{2}$. It is also found that the value of I is intimately related to the value of Z and $A-Z$ for the nucleus under consideration according to the following table:

Z	$A-Z$	I
Even	Even	0
Odd	Odd	Integer
Even	Odd	Half-integer
Odd	Even	Half-integer

Thus, when A is even the spin is zero or integer and when A is odd the spin is half-integer. We shall shortly see that these relations are one of the deciding features in distinguishing between different possible theories of the nucleus.

2.5 Magnetic moment

Since the nucleus is charged it is to be expected that if it is also spinning then, because a circulating charge is an electric current, it should have a magnetic moment. This is indeed found to be the case; all nuclei with I greater than zero have a magnetic moment whose value is measured in nuclear magnetons μ_0 where $\mu_0 = e\hbar/2Mc$ (M = proton mass, c = velocity of light).

2.6 Electric quadrupole moment

The quadrupole moment of a distribution of electric charge is a measure of the extent to which the charge distribution deviates from spherical symmetry. If a nucleus is ellipsoidal in shape, for instance, then the quadrupole moment Q is given by the expression

$$Q = \tfrac{4}{5} Z R^2 \frac{\Delta R}{R},$$

where R is the average nuclear radius and ΔR is the difference between the major and minor axes of the ellipsoid. Q can be both positive or negative corresponding to the ellipsoid being prolate or oblate (i.e. like a rugby football or a flying saucer) respectively. Quadrupole moments are found to vary considerably in size from almost zero to values corresponding to $\Delta R/R \simeq 0.4$. This is strikingly shown in Figure 2 where observed values of Q/R^2 for odd nuclei are plotted against Z or $A-Z$, whichever happens to be odd. The important fact is brought out by this diagram that Q changes sign for certain values of Z or $A-Z$, namely, 2, 8, 20, 28, 50, 82, 126. This fact is the first of many examples of the observation that nuclei for which Z or $A-Z$ are equal to one or other of these numbers have distinctive properties associated with them.

2.7 Radioactive decay

The fact that many nuclei are radioactive, that is, change from one nucleus to another or from one state to another, and at the same time emit radiation of some kind, is now commonplace knowledge. Three types of radiation are emitted, known as α-rays, β-rays and γ-rays. We shall consider first α- and β-rays which are both particle emissions, the former being helium nuclei

($Z = 2$, $A = 4$) and the latter electrons or positrons (a positron is a particle having the same mass as an electron but positively charged).

In α-decay it is always observed that the α-particles are emitted

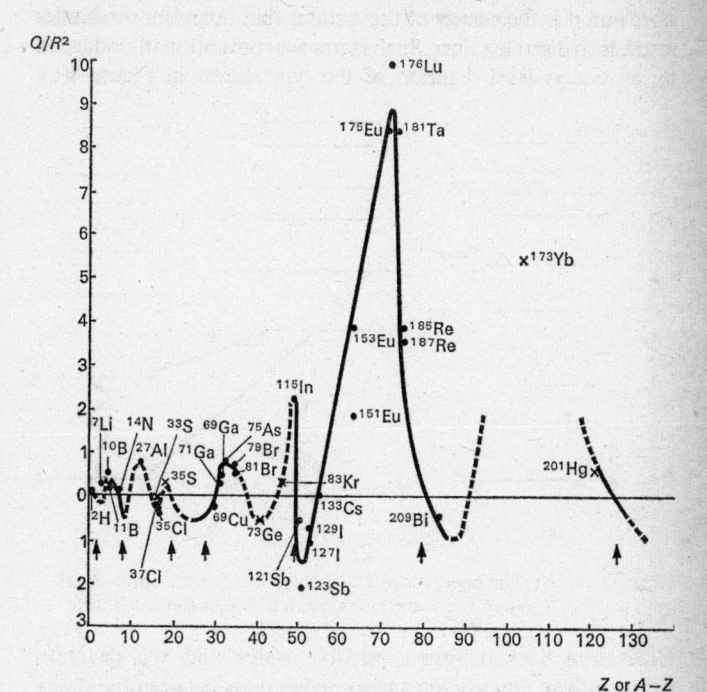

Figure 2 Experimental values of Q/R^2 for odd-A nuclei plotted as a function of Z or $A-Z$ whichever is odd (from Townes, Foley and Low, *Physical Reviews*, vol. 76, 1949, p. 1415)

either with one definite energy or sometimes in several groups, each group having a well-defined energy. This behaviour is very significant and can easily be understood in terms of quantum mechanics. We have already mentioned that quantum mechanics limits the possible values of the angular momentum of a system to be integer to half-integer multiples of $\hbar$; similarly it restricts the possible values that the energy of a system can take to discrete values. Normally a nucleus is in its state of lowest energy; however, just as a liquid can be given energy by heating it, so a nucleus can be given energy by, for example, bombarding it with other nuclear particles. The nucleus is then raised into an *excited state* and it is the energy of these states that quantum mechanics restricts to discrete values. Such states are conventionally indicated by an energy-level diagram of the type shown in Figure 3(a).

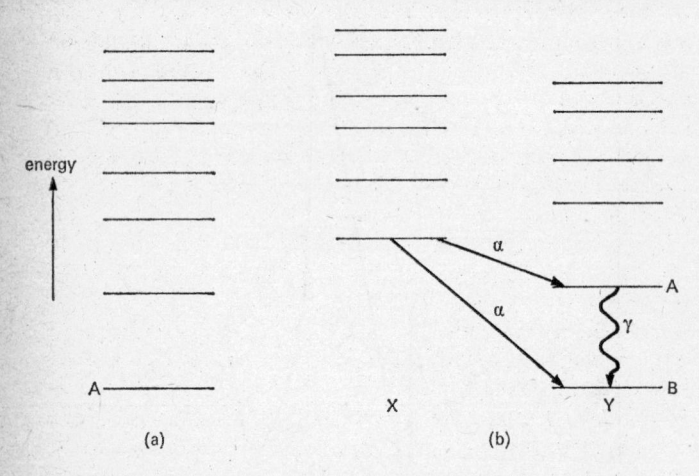

Figure 3 (a) Nuclear energy levels. A is the ground state. (b) α-decay from nucleus X to nucleus Y and γ transition between states A and B

Horizontal lines represent possible states and the distances between them indicate the energy separations between the states. A, the state of lowest energy, is referred to as the ground state and the remainder as excited states.

Now consider the α-decay of nucleus X to nucleus Y shown in Figure 3(b). Since the decay takes place at all it implies that X has more energy than Y, hence it is drawn higher in the diagram. If X was in its ground state then two possible transitions could be made; these are indicated by the two arrows. Since the transitions are between states of definite energy it is clear that since energy is conserved the α-particles must be emitted with two definite energies and, in observing a large number of decays of the X nuclei, the particles will be divided into two energetically different groups, the intensities of each group being proportional to the probabilities of the two transitions.

In the light of this argument it is then of considerable significance hat in β-decay the electrons (or positrons) are emitted with a spread of energies up to a certain maximum. How can this behaviour be reconciled with the foregoing discussion? The answer is that in β-decay *two* particles are in fact emitted, the electron (say) and a neutrino. The latter is a curious particle which has zero electric charge and is believed to have zero mass. (In spite of this, relativity theory still allows it to take away energy and momentum.) Because of its properties it is extremely difficult to detect and it is only within the last year or so that its presence has been directly confirmed by experiment; previously its existence was only inferred. Given such a particle, then the electron and neutrino will share the 'definite' energy released in a transition between two nuclear states and electrons will be observed with all energies up to the maximum possible in the particular transition considered.

In α- and β-emission, since electric charge is carried away from the nucleus, the charge Z of the nucleus must also change. This means that the final nucleus is completely different from the initial one. On the other hand, γ-rays are very short wavelength electromagnetic waves and although they carry away energy, they do not carry away any charge. In this case, then, there is a change in the state of the nucleus, but that is all. In Figure 3(b) a γ-transition from state A to B in nucleus Y is indicated by the wiggly line between A and B. Here again we expect the γ-ray to have a definite energy equal to the difference between that of states A and B; this is in agreement with experiment. The situation is completely analogous to that obtaining in the case of emission of light by atoms, where again there is quantization of energy levels and only certain wavelengths (or energies) are seen.

2.8 *Regularities*

It was pointed out earlier that electric quadrupole moments change sign at nuclei for which Z or $A-Z$ equals 2, 8, 20, 28, 50, 82, 126. There are many other indications that nuclei associated with these so-called 'magic numbers' have distinctive properties. Among them may be mentioned:

(a) Such nuclei have much higher cosmic abundances.
(b) Nuclear binding energies are greatest at magic numbers.
(c) The excitation energies of first excited states are greatest at magic numbers.

Perhaps the most striking 'magic' nucleus is ^{208}Pb which terminates three of the well-known radioactive series and which has $Z = 82$, $A-Z = 126$.

3 Components of the nucleus

In the same way that an atom is described in terms of a nucleus surrounded by a certain number of electrons, so we should like to describe the nucleus as a composite structure formed from a number of fundamental particles. In the early days, the most well-known particles were the proton (i.e. the hydrogen nucleus) and the electron. An obvious suggestion was then that a nucleus with atomic number Z and mass number A had as its constituents, A protons and $A-Z$ electrons. This ensures that the charge is Ze and that the mass is approximately correct (remember that the mass of an electron is nearly two thousand times smaller than that of a proton, so that the nuclear mass would be roughly that of A protons). Further, the emission of electrons in β-decay seems to support this hypothesis. However, there are two strong arguments against it.

Firstly, in quantum mechanics there is a very famous result known as Heisenberg's uncertainty relation. This relation tells us that if we confine a particle to a small region of space, then there is uncertainty as to its momentum; the particle acquires a 'zero-point' energy, a sort of violent wobbling whose magnitude depends on Planck's quantum constant h and is inversely proportional to the size of the region to which the particle is confined. A simple calculation shows that an electron confined to a volume of nuclear size would acquire a zero-point energy of the order 20 MeV† so that the Coulomb force between the protons and electrons, and this is the only force acting between them, would be quite unable to keep the electrons in the nucleus.

---

†1 MeV (million electronvolts) is the energy necessary to accelerate an electron through 1 million volts. It is a convenient unit for the measurement of nuclear energies and is equal to $1 \cdot 6 \times 10^{-13}$ J.

Secondly, both the proton and electron have spin $\frac{1}{2}\hbar$ and the rule for addition of angular momentum in quantum mechanics is that an odd number of half-integer spins gives a resultant half-integer spin, whereas an even number gives an integer spin. Now on the above picture for a nucleus (A, Z) the total number of particles (protons and electrons) is $A+(A-Z) = 2A-Z$. Thus the nucleus should have half-integer or integer spin according as $2A-Z$ is odd or even. This is manifestly in disagreement with observation which is that the spin is determined by whether A is odd or even (see section 2).

Thus on the above two counts the proton–electron picture of the nucleus is untenable. However, in 1932 Chadwick discovered the neutron. This is an electrically neutral spin $\frac{1}{2}\hbar$ particle having approximately the same mass as the proton and is clearly an excellent candidate as a nuclear building block. The nucleus can now be regarded as made up of Z protons and $N = A-Z$ neutrons. In this case the uncertainty relation does not give unreasonable results for the zero-point energy of the component particles since they are so much heavier than an electron. Further, the total number of particles in the nucleus is A so that the correct relationship between A and nuclear spin is obtained.

Of course, β-decay is now more difficult to understand since there are no longer any electrons in the nucleus. The current explanation of this is common to many similar processes involving elementary particles, namely that a neutron, for example, can *spontaneously* change into a proton together with an electron and a neutrino (symbolically n → p+e+ν). It is to be noticed here that apart from needing a neutrino to account for the energy spread of electrons in β-decay we also need it in the process to conserve angular momentum. For, if the neutron decay process had been n → p+e (without a neutrino) then the even number of particles (p+e) on the right-hand side, having integer spin, would not match the spin $\frac{1}{2}\hbar$ of the neutron. On the other hand, if the neutrino has spin $\frac{1}{2}\hbar$ (for which there is now considerable evidence) then angular momentum can be conserved.

4 Nuclear binding and nuclear forces

Given that the particles from which a nucleus is constructed are neutrons and protons, we have next to consider the way in which they are bound together. The magnitude of this binding can be estimated very easily for any given nucleus (A, Z) by comparing the actual mass of the nucleus $m(A, Z)$ with the sum of the masses of its constituent particles. That is we compare $m(A, Z)$ with $Zm_p + Nm_n$ where m_p and m_n are the masses of the proton

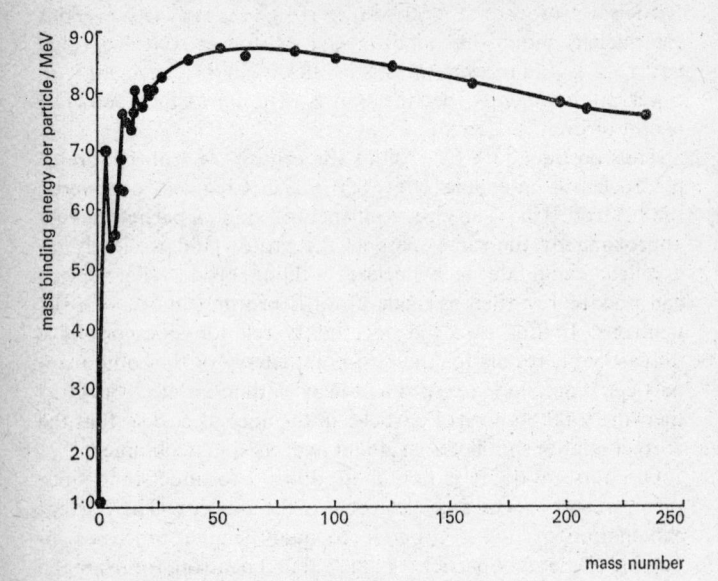

Figure 4 Binding energy per particle (in MeV) plotted against A

and neutron respectively. In the case of a stable nucleus $m(A, Z)$ is always less than the sum of the masses of its constituents. Bearing in mind Einstein's relation between mass and energy $E = mc^2$, it follows that energy has to be provided in some form in order to break up the nucleus and separate its components. Conversely, if we tried to build the nucleus up from neutrons and protons, then energy would be released in the process. The quantity $B = \{Zm_p + Nm_n - m(A, Z)\}c^2$ is known as the binding energy of the nucleus. A more significant quantity to discuss, however, is the binding energy per particle which is given by B/A. Its variation with A is shown in Figure 4 and a most important fact is demonstrated in this diagram, namely, that the binding energy per particle is approximately constant apart from a few very light nuclei. Its average value is about 8 MeV.

The obvious interpretation is to describe the binding of nuclei in terms of forces between nucleons (this is a collective name used to describe both neutrons and protons). These forces must be predominantly attractive in order to hold the nucleons together and furthermore must have a very short range of the order 10^{-13} cm so that nuclei have their observed sizes. Of course, the easiest way to determine the nature of these forces is to consider the interaction between only two nucleons. This can be effected by investigating the way in which neutrons and protons are scattered by each other and also the properties of the deuteron, the latter being the simplest possible nucleus consisting of a neutron and a proton only. Such investigations have been carried out and indicate quite definitely that nuclear forces are indeed of the required nature and that, in addition, they are charge independent. That is to say, they are essentially independent of whether the interacting particles are neutrons or protons. In addition, the experiments can be refined so that measurements are made of the way the forces behave when the spins of the nucleons are oriented in different directions relative both to each other and to their relative orbital angular momentum. There is clear evidence that the strength of the force depends on whether the spins of the two nucleons are parallel or anti-parallel to each other and it also seems highly probable that the force depends on how the spins are oriented with respect to the orbital angular momentum. The latter behaviour is referred to as spin–orbit coupling. One final property of the inter-nucleon interaction also emerges, namely that when two nucleons approach very closely to each other ($r < 0.4 \times 10^{-13}$ cm) then the force between them becomes highly repulsive.

This repulsive behaviour provides an explanation of two facts which have been presented about the nucleus. Firstly, the repulsion prevents all the nucleons in the nucleus from collapsing inwards to form a structure whose size would be essentially independent of the number of nucleons. Rather, we now have the situation that the nuclear volume should be proportional to the number, A, of nucleons and this requires that the nuclear radius should be proportional to $A^{\frac{1}{3}}$

in accord with observation. Further, detailed calculations of this effect show that the average separation of two nucleons in the nucleus is of the order $1 \cdot 5 - 2 \cdot 0 \times 10^{-13}$ cm, corresponding to nuclei radii of the observed order of magnitude. Because of the magnitude of this average separation between nucleons it follows that in a large nucleus any one nucleon interacts with a comparatively small number of its neighbours since the nuclear force only extends over a distance of the same order of magnitude as the separation. Thus the binding energy of the nucleus, which reflects the total attractive potential energy between nucleons, is proportional to the total number of particles; that is B/A is approximately constant as observed. If, on the other hand, the nucleons all collapsed inwards then each nucleon would interact with all the others so that the binding energy would be proportional to the total number of *pairs* of nucleons, that is to $A(A-1) \simeq A^2$ in complete disagreement with experiment.

Summarizing then, the nucleus is a many-body system consisting of Z protons and $N = A - Z$ neutrons interacting with each other through a force having a range of the order 10^{-13} cm which is predominantly attractive at large distances and repulsive at short distances. Furthermore, the force depends not only on the distance between two nucleons but also on the orientation of their spins both to each other and to their relative orbital angular momentum. The nucleus is indeed a highly complex structure.

5 Nuclear models

To obtain an exact theory of such a complicated structure is not possible – at least with our present mathematical techniques. Therefore, as with many other insoluble problems in physics, the approach adopted in order to give some theoretical account of the more detailed nuclear properties has been to devise models of the nucleus which may be regarded as crude approximations to the true state of affairs. Such models are suggested by known properties of the nucleus (e.g. the existence of magic numbers) and by other semi-classical considerations. Furthermore, a model must be amenable to theoretical treatment so that results can be calculated with it and compared with experiment. The success of a model is then judged by the extent to which theory and experiment are in agreement. In the following the two most successful nuclear models will be described, the shell model and the collective model. Both have had considerable success in accounting for various nuclear properties and have thrown a great deal of light on the real state of the nucleus.

5.1 *The nuclear shell model*

The idea of a shell model for the nucleus is at once suggested by the existence of magic numbers, that is by the existence of certain nuclei which are particularly stable and might even be approximately described as *inert*. In atomic theory inert elements (the noble gases) are well known as structures which are very stable and whose stability lies in the fact that certain shells of electrons in the atom have been filled. This 'filling' arises because electrons have spin $\frac{1}{2}\hbar$ and therefore obey the Pauli exclusion principle, a principle which requires that in any quantum-mechanical system no more than one electron can exist in any one state of motion (or quantum state). Now in an atom the various states arise because the electrons are moving in the central Coulomb field of the nucleus and they are labelled by quantum numbers, conventionally denoted by n, l, m_l and m_s. The relationship of these quantum numbers to physical concepts is roughly as follows.

In quantum mechanics, a material particle cannot in general be completely localized; the uncertainty relation referred to in section 3 is a statement of this fact. Thus, in an atom, an electron may be regarded as smeared out into a charge cloud around the nucleus. The way in which the density of the cloud varies with radial distance from the nucleus is then determined by the quantum number n, whilst its angular variation is dependent on l and m_l. Furthermore, in a given quantum state, the energy of the electron depends on n and l, whilst the magnitude of its orbital angular momentum about the nucleus is given approximately by $l\hbar$ and its direction is determined by m_l. The three quantum numbers n, l and m_l are restricted to integer values and for a state with energy $E(n, l)$, m_l can only take the $2l+1$ values $l, l-1,\ldots, -l$. Finally, m_s can take the values $\pm\frac{1}{2}$ corresponding to the two quantum-mechanically allowed spin orientations for an electron.

Electrons are said to be in the same shell if they have the same values of n and l, that is the same energy in the Coulomb field of the nucleus. Clearly, since for given n and l, m_l can take $2l+1$ values corresponding to $2l+1$ possible orientations of the electron orbit and two electrons ($m_s = \pm\frac{1}{2}$) can go into each of these states, then a filled shell contains $2(2l+1)$ electrons. Now the ground state of an atom is its state of lowest energy and is obtained in terms of the foregoing description by filling up each level

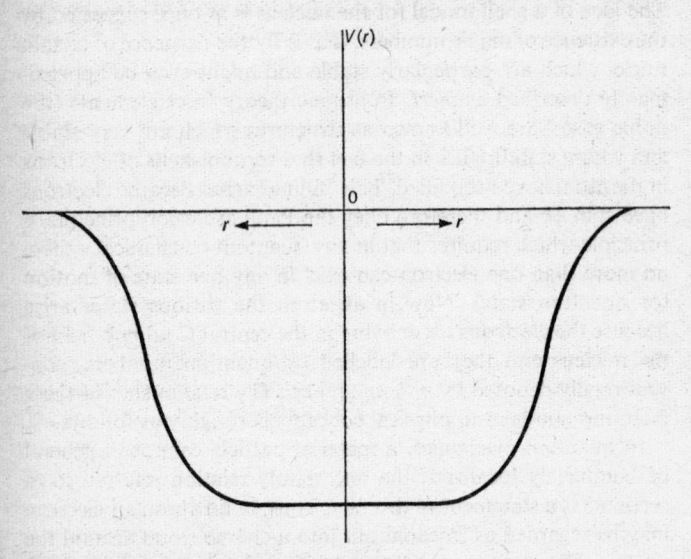

Figure 5 Nucleon potential energy $V(r)$

$E(n, l)$ in turn, starting from the lowest, until all Z electrons are accommodated. If as a result of this process the addition of the last electron just fills a shell, then the atom can be shown to be very stable and it is in this way the noble gases are accounted for.

One might hope for a similar explanation of the magic numbers which occur in the case of nuclei. However, whereas in the atom the different energy levels arise very simply, since the electrons are orbiting in the central Coulomb field of the nucleus, within the nucleus the situation is at first sight very different. For here it is not at all obvious how such a central field could arise, since there is no central massive particle to provide it. Nevertheless, a plausible explanation can be given.

Suppose we consider the potential energy $V(r)$ experienced by a nucleon as it moves from one side of a nucleus to the other. Away from the nucleus the potential is zero if we forget about Coulomb forces; as it approaches near to the nuclear surface it then experiences the attractive part of the potential due to the nearest nucleons so that the potential goes negative (Figure 5). Then well inside the nucleus the potential remains roughly constant on average at a certain negative value, since a given nucleon is only interacting with a few nucleons at a time and the situation is approximately independent of its position. Finally, emerging from the other side, the potential increases to zero again. Thus, as a first approximation we might regard the nucleons as moving independently of one another within a potential 'well' of the type shown in Figure 5, this well being the average field experienced by any one nucleon in the presence of all the others. In this well there will be a series of energy levels $E(n, l)$ whose relative spacings depend on the depth, size and shape of the well. The arrangement of these levels for such a well with a depth of about 40 MeV and radius 8×10^{-13} cm is shown on the left of Figure 6. In using the exclusion principle to control the filling of these levels it must be remembered that the principle only applies to *identical* particles so that a neutron in a level does not exclude a proton from occupying the level or vice versa. It can be seen that certain large gaps occur in the energy level scheme and that the total number of neutrons or protons that can be included below the lower gaps (remember $2(2l+1)$ particles can go into each level) correspond exactly with the magic numbers 2, 8, 20. However, the higher numbers do not appear.

It was not until 1948 that Mayer and Jensen showed how the higher numbers could be produced in a natural way. They pointed out that because of the spin dependence of nuclear forces, it is to be expected that a nucleon in the nucleus experiences a strong spin–orbit force such that its energy depends on whether its spin is parallel or anti-parallel to its orbital angular momentum $l\hbar$. That is, its energy is different according as its total angular momentum $(j\hbar)$, which is the vector sum of its spin and orbital angular momentum, is $(l+\tfrac{1}{2})\hbar$ or $(l-\tfrac{1}{2})\hbar$. This leads to a 'splitting' of the energy levels, and the energy of any level now depending on n, l and j and to agree with experiment the sign of the

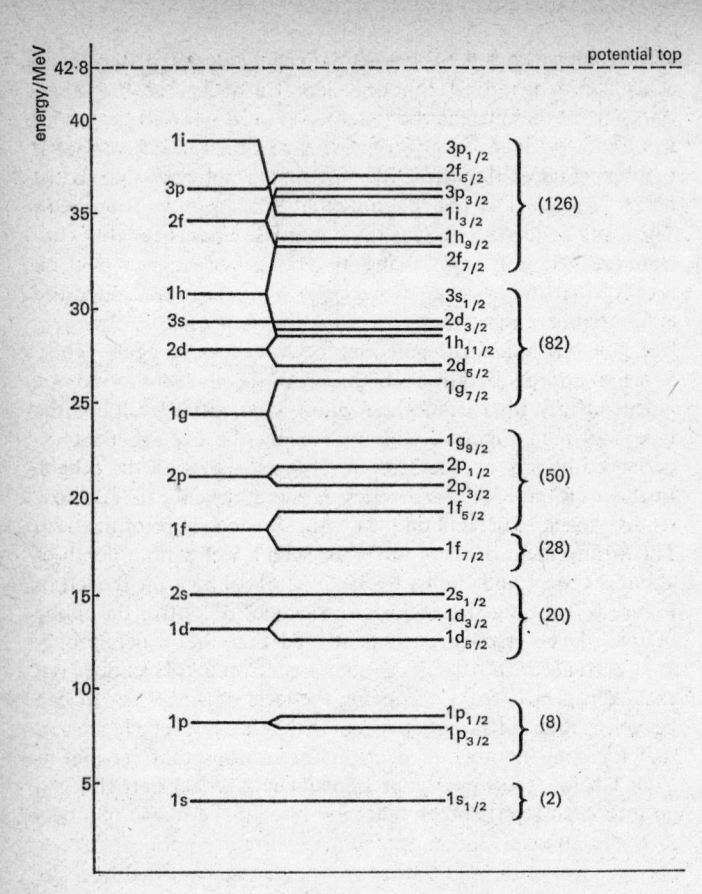

Figure 6 Energy levels in a potential well of the type shown in Figure 5. The notation used is as follows: States with $l = 0, 1, 2, 3,...$, are denoted by the letters s, p, d, f, g,..., The number to the left of each letter is the value of n and to the right, the value of j. On the extreme right of the diagram are the magic numbers representing the number of nucleons that completely fill all the preceding levels

spin–orbit coupling is such that the state of highest j lies lowest. Simple arguments can be used to show that this splitting should be approximately proportional to $2l+1$ leading to very large splittings for states of high l. The level scheme obtained because of this effect is shown on the right-hand side of Figure 6. Bearing in mind that because of the exclusion principle a level j can be occupied by a maximum of $2j+1$ particles it can be seen that all the magic numbers now occur in a natural way immediately below relatively large gaps. Of course, the larger a gap, the more stable is the immediately preceding nucleus since it requires more energy to excite it. What has happened as a result of introducing the spin–orbit coupling is that for the higher groups of levels the splitting of the state with the highest orbital angular momentum is so great that the lower component of the resulting doublet joins the next lowest group of levels.

We now have a basic model and next have to consider to what extent it can account for other nuclear properties. Its success, in fact, is phenomenal. Consider first of all the question of nuclear spins. Here we have to take account of the fact that the nucleons do not move completely independently of one another. Although the forces between the nucleons are partially accounted for by the average potential well that has been introduced, there are still residual nuclear interactions representing the difference between the well potential energy and that actually experienced by a nucleon. From the known nature of nuclear forces it is to be expected that these interactions are such that like nucleons in the different levels tend to pair off with one another, so that their total angular momentum is zero. Under these circumstances there is no contribution to the angular momentum of a nucleus having an even number of neutrons and an even number of protons since their orbits and spins are all paired off. This is in agreement with the observation that the so-called even–even nuclei have spin zero (see section 2). However, for an odd-A nucleus in the last neutron or proton energy level to be filled there is either an odd neutron or an odd proton which is unpaired. This is the only particle contributing to the nuclear spin, which should therefore be equal to the angular momentum $j\hbar$ of this nucleon. Thus ^{17}O has 8 protons and 9 neutrons; the last odd neutron has $j = \frac{5}{2}$ and this is the observed spin of ^{17}O. Similarly ^{59}Co has 27 protons and 32 neutrons; the last odd proton has $j = \frac{7}{2}$ which is again in agreement with the experimental spin value. It is, however, significant that in the cases where high spins are to be expected (e.g. when the $j = \frac{11}{2}$ state is being filled) they are not observed. The reason for this is that the residual pairing forces are much more effective in states of high angular momentum so that particles like to pair off in such states, even at the expense (energetically) of lifting a nucleon out of a lower angular momentum state in order to do this. In this case the nucleus has its spin equal to the lower angular momentum. Taking this into account, there are only very few odd-A nuclei whose spins are in disagreement with this simple shell model prediction, and in all such cases there are extenuating circumstances. As far as odd–odd nuclei are concerned (i.e. those with odd Z and odd N) the shell model in the form so far described makes no prediction other than that the spin should be integer.

The shell model also predicts the parities of nuclei. Parity has been a much discussed concept recently; roughly speaking it is a description of the way the nuclear properties change when investigated in a mirror world, that is, in a world in which the three coordinate axes have been reflected through the origin. It is sufficient for our purposes to note that it is said to be either 'even' or 'odd' according as the nuclear wave function remains the same or changes sign under reflection of axes. On the shell model the parity of an odd-A nucleus is dependent on the l-value for the last odd nucleon, being odd if l is odd and even if l is even. Again theory and experiment agree rather well.

Perhaps one of the most striking successes of the shell model is its ability to describe the general behaviour of the nuclear magnetic moments of odd-A nuclei. For these nuclei the magnetic moment is predicted to be that of the last odd neutron or proton as the case may be. This is because the contributions to the magnetic moment from the other nucleons exactly cancel due to the pairing effect. The theoretical and experimental results are presented in the so-called Schmidt diagrams (Figures 7 and 8) for odd-neutron and odd-proton nuclei. On the horizontal axis is plotted the nuclear spin and on the vertical axis the magnetic moment. The full lines correspond to the theoretical predictions for the two possibilities $j = l \pm \frac{1}{2}$ and of course only have any meaning at the allowed half-integer spin values at which are plotted the experimental points. The qualitative agreement is surprisingly good and can be considerably improved by refining the shell model in various ways.

When we come to consider nuclear electric quadrupole moments Q, however, there is violent disagreement. This arises mainly in the approximate regions $150 < A < 185$, $225 < A$ where the experimentally observed values of Q are sometimes as much as

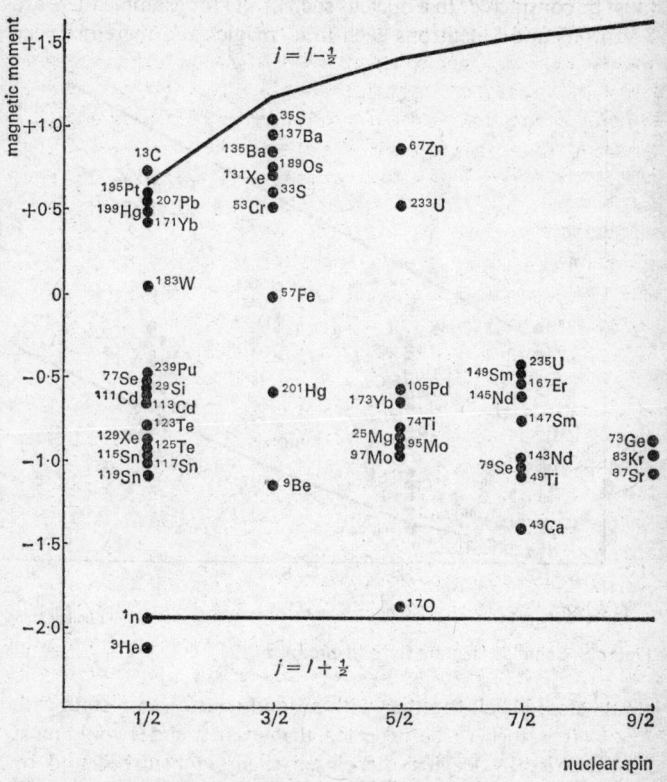

Figure 7 Schmidt diagram for odd-neutron nuclei

30 times larger than the shell-model predictions. These predictions are calculated in the same way as the magnetic moments by attributing Q to the last odd particle. In particular, since a neutron has zero charge, an odd-neutron nucleus is expected to have $Q = 0$; on the contrary this is never observed. The fact that quadrupole moments having many times the single-particle

values occur is significant since it indicates that many particles are contributing and that we have here a collective effect of some kind. This will be discussed in detail a little later.

Of the other nuclear properties, the nature of excited states must be considered. In a nucleus such as ^{17}O for example there are 8 protons and 8 neutrons each in a 'magic' and therefore very

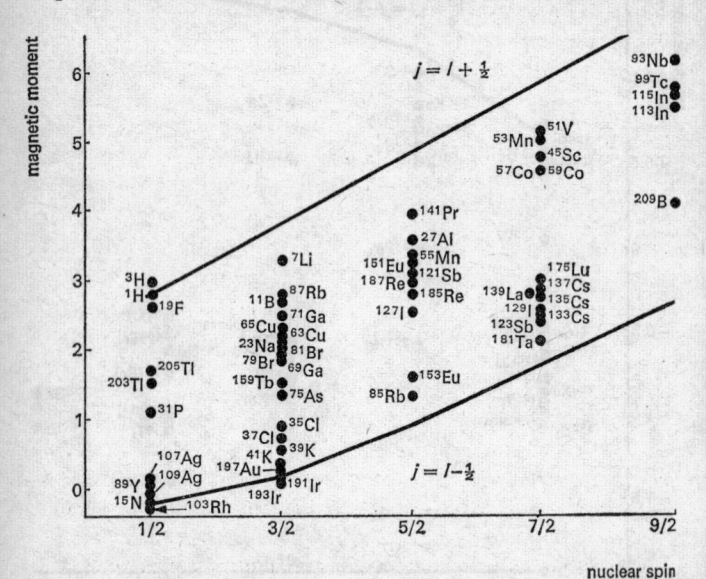

Figure 8 Schmidt diagram for odd-proton nuclei

stable closed shell together with an odd neutron in a state with $j = \frac{5}{2}$. It is then to be expected that excited states could most simply arise by leaving the closed shells undisturbed and by exciting the odd neutron into one of the higher states; thus among the low excited states we should expect one with $j = \frac{3}{2}$ and one with $j = \frac{1}{2}$. This is in agreement with experiment and there are many cases such as this where the low-lying excited states of nuclei can be interpreted in a very simple fashion. However, the case quoted was ideal and unambiguous. In a more complicated nuclei away from 'magic' closed shells the energy-level scheme cannot be deduced so simply and recourse must be made to detailed calculations taking into account the residual interactions between nucleons, the energy levels then being interpreted as re-arrangements of the nucleons within incomplete shells. Where it has been possible to carry out such calculations the theoretical results have been found to be in good agreement with experiment. Right away from closed shells, however, the necessary calculations are unfeasible and it is just in these regions that the collective model suggested by Aage Bohr has had some of its successes.

Before going on to describe the collective model, one final point has to be discussed, namely, why the shell model works so well. The basis of the shell model is the assumption that the nucleons move approximately independently of one another in a smooth potential well. Plausibility arguments were given for this assumption; but these arguments completely ignored the fact that since nuclear forces are so strong and have such a short range, a nucleon moving through the nucleus would in fact experience violent fluctuations in potential energy as it passed near other nucleons, in complete contradiction with our assumption of roughly independent particle motion. The presence of such effects is confirmed in a great many nuclear reactions in which a neutron, for example, strikes a nucleus and is immediately captured to form a highly excited state of the nucleus. Such capture would not take place if the neutron moved through the nucleus in the way suggested by the shell model since motion independent of the other particles implies that the nucleus is approximately 'transparent'.

The reason why, in spite of this apparently contrary evidence, the shell model might still be expected to work, was first given by Weisskopf. He pointed out that for nucleons deep down in the potential well the violent fluctuations are quite ineffective. This is because the only effect such fluctuations can have is to 'knock' a nucleon from one energy state to another nearby. However, for most particles all the nearby states are already occupied so that the Pauli exclusion principle forbids such changes of state. Thus for the bulk of the nucleons the violent changes in potential have little effect. On the other hand, in the case of a nuclear reaction, the incoming particle has an energy at least 8 MeV greater than that of any nucleon in the nucleus since nucleons only fill the potential well to within 8 MeV of the top. This means that it has sufficient energy to knock particles, even deep down in the potential well, up into unoccupied states. The incoming particle therefore interacts strongly with the nucleus although the bulk of the constituent nucleons of the nucleus move approximately independent of one another.

This explanation is, of course, a considerable simplification but it is essentially correct and is confirmed by the very detailed calculations that have been performed on this problem during the past few years.

5.2 *The collective model*

It was pointed out in the previous section that the quadrupole moments of a number of nuclei are much larger than the single-

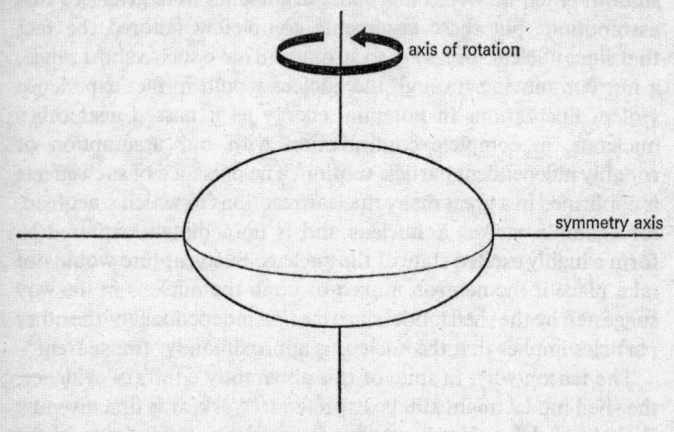

Figure 9 Rotation of an ellipsoidal nucleus

particle values predicted by the shell model. These nuclei are all those which are well away from closed shells and in which there is a large number of 'loose' particles, some of which have high angular momenta. Such particles will exert a high centrifugal force on the surface of the nucleus tending to distort its shape from approximately spherical to much more ellipsoidal, thus giving it a large quadrupole moment. This effect is collective in nature in that a great many nucleons participate in the overall nuclear distortion.

In the case of nuclei with such large distortions it might be expected that excited states of a rotational character should be observed. These states would be similar to those which arise in the case of diatomic molecules and are states of motion in which the nucleus rotates as a whole about an axis perpendicular to its axis of symmetry (Figure 9). Of course we must remember the limitations imposed by quantum mechanics on such rotations. Classically, the energy is given by $P^2/2\mathfrak{J}$ where $\mathfrak{J}$ is the moment of inertia of the nucleus about the axis of rotation and P the angular momentum. Quantum mechanics however restricts P^2 to taking the values $P^2 = J(J+1)\hbar^2$ where, for an even–even nucleus, J takes even integer values only. For odd-A nuclei a similar but more complicated description can be given but we shall not consider these nuclei here. Thus, for even–even nuclei the allowed rotational energy levels are given by

$$E_J = \frac{J(J+1)\hbar^2}{2\mathfrak{J}},$$

with $J = 0, 2, 4,\ldots$.

It is a remarkable success of such a simple theory that energy levels agreeing very well with this formula are observed just where they are to be expected (the regions $150 < A < 185$, $A > 225$). A typical example is shown in Figure 10. An interesting feature of rotational energy levels of this kind is that the moment of inertia $\mathfrak{J}$ deduced from the level spacing is roughly equal to one-half the moment of inertia that would be obtained if the nucleus rotated rigidly. This means that the nucleus behaves rather like a drop of liquid and that the rotation can be pictured as taking place somewhat as in Figure 11, the central region remaining stationary and the external region moving round like a wave.

Collective motion of nuclear matter can also be obtained in regions where Q is not so large in the form of harmonic vibrations about the equilibrium shape. This sort of motion is characterized by a series of equally spaced energy levels, the spacing being $h\nu$ where ν is the natural period of oscillation of the nucleus. Again it is satisfying that energy levels of this type have been observed for a number of nuclei. In fact, there appears to be an abrupt transition at $N = 88$ in that for nuclei with $N > 88$ the characteristic level spacing for rotational levels is observed, whilst for $N < 88$ the excited states are vibrational in nature.

These vibrational levels can also be described in terms of the vibration of a drop of liquid, in the same way that rotational

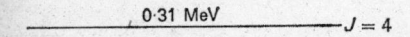

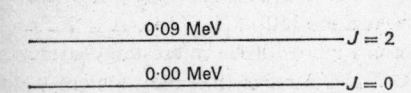

———————— 1·09 MeV ———————— $J = 8$

———————— 0·64 MeV ———————— $J = 6$

———————— 0·31 MeV ———————— $J = 4$

———————— 0·09 MeV ———————— $J = 2$
———————— 0·00 MeV ———————— $J = 0$

Figure 10 Rotational energy levels in ^{180}Hf

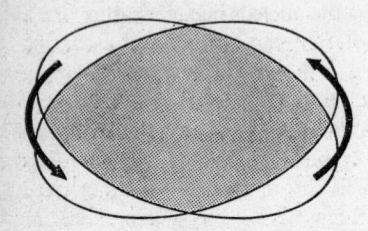

Figure 11 Qualitative diagram representing the motion of nuclear matter during the rotation of an ellipsoidal nucleus

levels can be related in some sense to the rotation of a distorted liquid drop. The idea that the nucleus might be compared to a liquid drop is, of course, not new; it was first suggested by Niels

Bohr in 1936 in connection with a description of nuclear reactions. A nucleon entering a nucleus and giving up its energy is likened to a heating up of the drop and the subsequent emission of particles in the nuclear reaction is analogous to evaporation. The liquid-drop description of the nucleus is also strikingly supported by the fission process in which a nucleus struck by a neutron undergoes such a violent 'shaking' that it breaks up into two large nuclear fragments. This simple description cannot, however, account for any very detailed properties of the nucleus; in such cases reference always has to be made to individual nucleons in the nucleus.

6 Conclusions

The shell model and the collective model are the two basic models of the nucleus and between them they account in a semi-quantitative fashion for a great many nuclear properties. The obvious question to ask is to what extent the models are equivalent and whether they approach a 'true' description of the nucleus.

The problem of their equivalence has occupied theoretical physicists for a number of years. Recently some success has been achieved in showing that shell-model states can on occasions closely resemble the rotational and vibrational states of the collective model and it has even been possible to make approximate calculations, for example, of the moment of inertia or the spacing between vibrational levels.

Another approach to these problems has been to unify the ideas of both models in the so-called 'unified model'. In this treatment is it supposed that the nucleons are moving in an ellipsoidal potential well rather than a spherical one and that the whole can undergo collective rotational motion. The level schemes then become very complicated, but it has nevertheless been found that this description does help to account for a large number of experimental data.

All the models to some extent illuminate the true nature of the nucleus and are at the moment the most suggestive and useful way of describing it. However, there have recently been a number of developments in techniques for the direct mathematical treatment of the formidable problem presented by the many-body nature of the nucleus. Such calculations have as one of their main objectives the prediction of the various parameters that we have already introduced empirically in the various models such as the nuclear size, density distribution, binding energy, potential-well depth, etc. Furthermore, the detailed theory must also show the relationship between the true nuclear state and the corresponding model state. Already some results have been obtained on the nuclear size and binding energy and it seems clear that within the next few years the detailed calculations will be improved enormously giving us very considerable information about the actual state of the atomic nucleus.

6 F. R. Stannard
High-Energy Nuclear Physics I. Experimental Techniques

F. R. Stannard, 'High energy nuclear physics, I. Experimental techniques', *Physics Education*, vol. 1, 1966, no. 3, pp. 154–62.

What the scientist really desires is for his science to be understood and to become an integral part of our general culture.
Rabi, *My Life and Times as a Physicist*.

On the outskirts of Geneva is CERN, the European centre for nuclear physics research. A sprawling complex of buildings clusters around an intriguing vast mound of earth. Shaped like a doughnut, 180 m in diameter, this mound serves to protect CERN's two thousand workers from the radiation given off by the giant particle accelerator beneath it. This atom smasher, along with others to be found throughout the USA, Russia, Britain and Europe, is the tool which high-energy physicists use to investigate the fundamental laws of nature and the constitution of matter itself.

It is customary these days to speak of 'little science' and 'big science'. The former includes research in biology, chemistry, etc., whereas the latter embraces space exploration and high-energy physics. Big science becomes increasingly more expensive year by year. CERN's annual budget alone comes to £10 million. These new accelerators being planned will cost over £130 million each, and contracts for building them are the subject of negotiation between governments. Clearly, expenditure on such a scale can only be justified for work of the highest interest and potentiality. What is it all about?

The equivalence of mass and energy

We begin with Einstein's famous equation relating the mass m of a system to its equivalent energy E,

$E = mc^2$,

where c is the velocity of light. The important point is that energy and matter are merely two different manifestations of the same phenomenon. One must no longer talk of the law of conservation of mass or the law of conservation of energy, but rather of a more general law which requires the total sum of mass plus energy to remain constant. As long as this condition is satified, mass may transform into energy and vice versa.

One all too familiar example of such a transformation is the nuclear bomb. In such a device one has a quantity of radioactive material undergoing nuclear reactions. These reactions are such that if one could gather up all the pieces after the explosion some mass would be missing. Einstein's equation specifies how much energy should be released as a result of a given reduction in mass. It happens that what in normal everyday life would be considered to be a very little mass is equivalent to a great deal of energy. Thus, in a nuclear bomb such as that exploded at Hiroshima, the enormous energy release resulted from a reduction in the mass of the original radioactive material by only one gramme.

High-energy physics research is not so much concerned with the conversion of mass into energy as with the reverse process – the creation of matter and the study of its subsequent behaviour. Einstein's equation, which acts so advantageously when making nuclear bombs or nuclear power stations, works unfavourably in this field. Vast quantities of energy are needed to produce even the most modest amounts of new material.

Cosmic rays

The basic process used to create matter is to take a nuclear particle, give it a great deal of kinetic energy, and then make it strike a second nuclear particle. At the moment of impact, in a manner that is far from understood, some of the kinetic energy may convert into mass. Thus, coming away from the collision are three, four or more particles.

So one's first need is for an energetic particle. In this respect nature is kind, and provides a free source of such projectiles – cosmic rays. These consist of nuclear particles, mostly protons, coming from outer space with enormous velocities. On approaching the Earth they strike the nuclei belonging to the atoms of the atmosphere and send out showers of new particles. These in turn make collisions, gradually slow down and stop; a few get through to the Earth's surface. At sea-level practically all that can be observed is the 'debris' from the more interesting phenomena occurring at greater altitudes. Much is therefore to be gained by taking one's detecting equipment up mountains or flying it in balloons. English physicists in particular were quick to seize on this point; for, living on an island virtually devoid of mountains, the prospects for foreign travel and skiing suddenly brightened.

Many happy and exceedingly fruitful years were spent in this manner. But after a while it became clear that waiting for the occasional cosmic ray to enter one's equipment was a most haphazard and unsatisfactory procedure. A stage was reached where little further progress could be made unless one could observe collisions in numbers many many orders of magnitude greater than were ever likely to be achieved with cosmic rays. A further drawback was that the exact energy of the cosmic-ray particle was never known and this severely hampered the interpretation of the 'event', i.e. the collision.

Thus, in the early 1950s interest was diverted towards producing one's own projectiles. The object was to accelerate large numbers of protons from rest up to a carefully specified kinetic energy and then observe their collisions with nuclei. Although most high-energy physicists made the change over to accelerators, one should not lose sight of the fact that cosmic rays are still of great interest, especially when a particle arrives with what one might call a super-high energy. These energies are far beyond the reach of any accelerator that can be envisaged. Consequently, some physicists continue their studies of cosmic rays, specializing in these vary rare spectacular events.

Accelerators

The principle behind an accelerator is that a proton acquires kinetic energy when placed in an electric field, due to the interaction between its charge and the field. The energy obtained depends upon the potential difference through which it moves, so this leads one to define a basic unit of energy – the electronvolt. This is simply the energy acquired by any particle carrying a charge equal in magnitude to that of the charge on an electron, when it traverses a potential difference of one volt. In terms of more familiar units, one electronvolt (eV) is equal to 1.6×10^{-19} joules. In practice one finds that about 200 million are needed to produce even the smallest new particle, so as a unit, the electronvolt is rather tiny for our purposes. Hence, one further defines an MeV as a million electronvolts, and a GeV as 1000 MeV (in the USA one thousand million is a billion, so this latter unit has been called a BeV). Incidentally, in terms of energy the mass of an electron is equivalent to 0.51 MeV and that of the proton to 938 MeV.

Early accelerators, like the Van de Graaff, depended upon building up the highest possible voltage difference, and then accelerating the particle from one electrode to the other. Unfortunately, owing to the problem of providing adequate

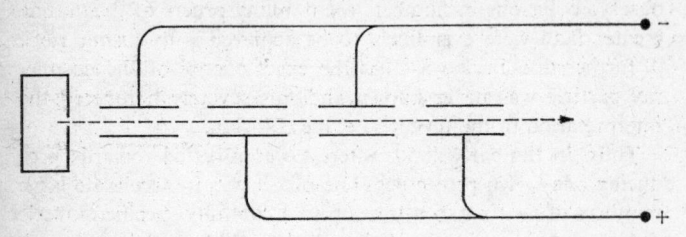

Figure 1

insulation, seven million volts is about the absolute maximum one can hope to achieve, and a 7 MeV proton is of little use to us.

At this stage one has to use a little cunning. It is to be noted that whereas a charged particle will accelerate when placed between two electrodes, if it is inside a conductor its motion is unaffected by changes of potential on the conductor. Thus, in the arrangement shown in Figure 1, there are a succession of electrodes, cylindrical in shape. A proton from the source is attracted across the first gap and enters the first cylindrical electrode. Once it is inside, the polarity of all the electrodes is switched so that, after drifting with uniform velocity for a time, the particle on emerging into the next gap is presented with a second accelerating voltage. This process can be repeated indefinitely with the energy of the particle being built up in small steps.

Eventually, of course, one comes across difficulties of a financial character which limit the number of electrodes one can have. As might be expected, the longest accelerator in the world is being built in the USA at Stanford. This monster when completed will be two miles [3 km] long and will accelerate electrons to an energy of 50 GeV. One of its more amusing statistics is that the annual rainfall at its two ends differ by two inches [5 cm].

In order to reduce costs, and contain the accelerator on a more modestly sized site, a little more ingenuity is required. So one next makes use of the fact that the motion of a charged particle can be altered by a magnetic, as well as an electric, field. We are all familiar with the idea that a wire carrying an electric current, when placed in a magnetic field which is at right angles to its length, experiences a force at right angles to both its length and the field. A current is merely electrons in motion, so it is not surprising that all moving charged particles show the same effect. Acting at right angles to the motion, the force is used to change the particle's direction, not its speed.

The electrodes, instead of lying in a straight line, are now arranged in a large horizontal circle. A vertical magnetic field serves to steer the protons round the circle until eventually the particles pass for a second time through the same electrodes. The process can be repeated for many circuits with the particles increasing in energy at every revolution. This is the principle of the *proton synchrotron*, which is the main type of powerful accelerator used these days. In synchrotrons the final energy of the protons in a given machine is restricted by the maximum attainable magnetic field. As the particles accelerate so they become more and more difficult to bend into the arc of a circle. The limit is reached when the proton's momentum is such that it requires the maximum field to keep it on the desired course. To reach higher energies a larger radius is necessary, and hence we see the reason for the spectacular dimensions of modern-day accelerators. The CERN and Brookhaven synchrotrons are the largest in the world and produce protons of 30 GeV.

These machines, besides being large, are of extraordinary complexity and precision. For example, as the particles accelerate the frequency of the accelerating voltage must be carefully increased also, otherwise the particles would get out of step. The magnetic field must be progressively increased as well during the acceleration period so that although the particle's momentum changes it still travels along the same path. These and other aspects of accelerator operation are governed from [the] main control room – a room that appears to belong to the realms of science fiction, with close to one hundred 2 m high racks of electronics and television screens. The magnet itself has to be very accurately manufactured and mounted. The CERN proton synchrotron magnet consists of one hundred separate sections, each weighing 30 tonnes, and these have to be aligned to an accuracy of tenths of a millimetre. This is because the protons during their second acceleration period make half a million circuits and travel a distance equivalent to going round the Earth several times. It would only need a slight misalignment of one of the magnet sections to throw the proton completely off course.

In order to protect the operators and physicists from the harmful radiation that is generated by accelerators these machines are contained in massive circular buildings made of concrete and covered with earth. All operation of the accelerator is carried on by remote control. Elaborate precautions are taken to ensure that no one by mistake strays into the danger area while the machine is operating. For example, if a technician needs to repair some fault he will remove a key from a panel in the main control room and, in so doing, automatically switch off the accelerator. This key is necessary to unlock the door to get at the accelerator. While inside the technician knows he can work in perfect safety for the machine cannot be restarted until he has relocked the door and returned the key to its position on the panel in the control room. A further precaution is provided by light beams which criss-cross the danger area and, if interrupted by someone walking about, stop the machine. There are, in the interests of safety, so many ways of stopping the accelerator that it is a wonder the machine ever goes!

One perhaps should remark at this point that when a charged particle moves in a circle it sprays out electromagnetic radiation and so loses some of the energy it has gained from the accelerating field. For electrons, which are so much lighter than protons, the effect eventually becomes so serious that at each revolution they lose as much energy by radiation as they have gained. Hence, very energetic electrons can only be obtained if the acceleration takes place in a straight line, and this is the reason for building the straight (or linear) electron accelerator at Stanford, rather than a synchrotron.

And what of future proton accelerators? Clearly, one approach is to build bigger conventional synchrotrons; the present plan is to construct one with a diameter of $2\frac{1}{2}$ km, capable of producing 300 GeV protons. An additional scheme is to exploit the 'clashing beam' principle. To appreciate the idea behind this, one must first understand what the physicist means when he talks of 'the centre-of-mass system'. Let us consider a proton striking a stationary target nucleus. Besides having kinetic energy, the moving particle also possesses momentum and this, by the law of conservation of momentum, must be transferred to the final set of particles. But if the particles leaving the collision have to possess momentum, then they too must have some kinetic energy. Thus, one cannot use all the kinetic energy of the incoming proton for producing new particles; some must be held in reserve and given to the final products. The energy that actually is available for producing new particles is called 'the energy in the centre-of-mass system', or the *c.m. energy*. It is here that nature deals a cruel blow. It turns out that the higher the energy of the proton the smaller is the percentage of the energy that is actually of use. For example, if a proton of 1 GeV strikes a stationary proton, the c.m. energy is 0·43 GeV; at 6 GeV only 2 GeV is available, at 10 GeV, 2·9 GeV, and at 100 GeV, 10·5 GeV. The clashing beam principle is an idea for circumventing this difficulty. As the name implies, one takes two beams of protons travelling in opposite directions and makes them collide head-on. The two colliding protons can be arranged to have equal and opposite momentum, so the total overall momentum of the system is zero, and *all* the energy of the protons is c.m. energy. Thus clashing beams of 5·25 GeV have a c.m. energy of 10·5 GeV which, as we have just seen, is as much energy as one can get from a conventional 100 GeV synchrotron. There is, however, at least one reason why machines of this type will not supplant the conventional ones. This is the fact that, having created new particles, one wishes to observe their behaviour when they in turn collide with nuclei at various particle energies. With a conventional synchrotron the new particles are created with a wide range of energies, whereas with the clashing beam machine the range of energies with which the new particles are emitted from the collision has been deliberately restricted. Nevertheless, clashing beams provide the physicist with an exciting glimpse of what lies in store for him when he manages to reach higher energies by the usual method.

As for the far-off future, one may idly speculate on what archaeologists thousands of years hence will make of the large submerged concrete circles. No doubt, many a scholarly treatise will be written drawing comparisons between them and Stonehenge. If the remains of the 30-tonne cranes are missing these experts will even have plenty of scope for guessing how we managed to get one block of concrete on to another.

Beams

The circulating protons having been accelerated to full energy, the next stage is to flick up a metallic target to intercept them. The resulting collisions between the protons and the nuclei of the metal then lead to a shower of protons, neutrons, aggregates of protons and neutrons, and new particles going in all directions with a wide range of energies. The physicist, however, is not satisfied with just producing the new particles. He would like to investigate their behaviour to the same extent as he does that of the proton, and this entails somehow producing a beam consisting of one new particle of a given mass and electric charge, travelling with a specified velocity. The problem is, therefore, how to sort out from the shower coming from the target the particular particles of interest.

The procedure calls for a high degree of sophistication, but we can describe the basic elements with the aid of Figure 2. In region A are negatively and positively charged particles of all energies, masses and directions. The first stage in the refining process is to eliminate all particles except those travelling in some given direction. This is done by massive concrete shielding blocks which have a narrow slit in them. On emerging from the slit the particles pass into a magnetic field, which in the diagram is directed into the paper. Depending upon the sign of its electric charge, a particle is deflected to the left or to the right. Moreover, as already noted, the *amount* of the deflection depends upon the particle's *momentum*; the smaller the momentum, the greater the deflection. Thus, emerging from region B through a second slit, one has particles of one electric charge, one momentum, one direction, but still with a wide variety of mass.

In region C the particles pass through an electric field and are deflected once again. But here, because the electric charges on the particles are equal, the force experienced by all the particles is

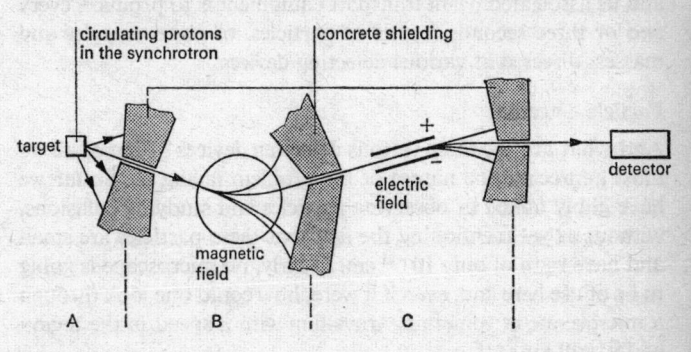

Figure 2

the same. Hence the amount of the deflection depends solely on the length of time each particle spends in the field, and consequently the separation is now according to *velocity*. But the particles have already been selected to have the same momentum. Recalling that

Momentum = mass × velocity,

one immediately sees that this last step in effect separates according to mass. Thus emerging from region C are particles of unique charge, mass, momentum and direction. These are then led into a detector of some kind and their collisions with nuclei studied.

By altering the strengths of the magnetic and electric fields, and the positions of the concrete shielding blocks, one can select different particles and energies as desired.

It might be thought that this was a very wasteful procedure, for it is clear that at any given time one is studying only a tiny fraction of the particles produced. However, it should be borne in mind that the machine accelerates one million million protons each period, so one can afford to throw away some of them! Nevertheless, great care is taken to use the machine as efficiently as the layout of concrete blocks, magnets, etc., will allow. With the complex arrays that surround the accelerator, several beams can be taken simultaneously off the same target, and indeed the protons are made to strike more than one target.

Summing up we can say that the function of an accelerator and its associated beam transport equipment is to produce, every two or three seconds, bursts of particles, of given energies and masses, directed at various detecting devices.

Particle detectors

And what are the these various detecting devices? First of all we must appreciate the nature of the problem facing us. So far we have glibly talked of observing particles and studying collisions, without as yet mentioning the fact that these particles are small and have radii of only 10^{-13} cm. Clearly, no microscope is going to be of use here and, even if it were, how could one look through a microscope at something travelling with a speed in the region of 150 000 km s^{-1}!

The situation would have been hopeless, and high-energy physics research would never have got started, had it not been for a phenomenon called *ionization*. Ionization is the key process for making the particles, or more strictly speaking their trajectories, visible.

To understand what it is about we must first remember how an atom is constructed. At the centre is a nucleus which itself is made up of protons and neutrons bound tightly together. Depending on the number of protons present, so one can make up nuclei belonging to atoms of different elements. Around the nucleus revolve electrons much as the planets go round the sun in the solar system. The electrons are very light, having about 1/1836 of the mass of the proton. In a normal atom the number of electrons in orbit is the same as the number of protons. Since the protons have one unit of positive charge, the electrons one unit of negative charge, and the neutrons are electrically neutral, the overall charge of the atom is zero.

When we have talked of protons making collisions we were referring to collisions on the central nucleus directly. The chances of this happening, however, are not good, as we can see if we imagine an atom expanded to the size of London Airport. On this scale a proton would be no bigger than a golf ball. The reader is left to work out for himself the chances of hitting it with a second golf ball dropped from an aeroplane! Notwithstanding, a proton, or any other charged particle, can still do a great deal of damage to an atom without going anywhere near the nucleus. This arises because, as is well known, two charged bodies exert a force on each other, no matter what their separation. The projectile consequently has only to pass reasonably close to an electron to influence it to a marked extent. Because of its small mass the electron is easily pushed around and often finds itself knocked clean out of an atom. An atom so denuded of one of its electrons has an overall positive charge and is said to be ionized. The projectile carries on virtually unaffected, ionizing further atoms along its way until eventually it scores a direct hit on a nucleus. The charged particles emitted from the collision can themselves go on to ionize atoms as they move through the medium.

Instead of trying to detect a tiny particle travelling with enormous speed, the problem now simplifies itself into one of finding methods for detecting the presence of the tell-tale trail of ionized atoms left in the particle's wake. What happens in a particle detector bears some similarity to the growth of an avalanche. At the start a small stone becomes dislodged near the summit of the mountain. As it falls, rocks, snow and ice follow, and gradually the whole process builds up to enormous proportions, but in order for it to happen at all the snow and rocks on the mountainside have to be inherently unstable, needing only the slightest push in order to allow the gravitational forces to take effect. So it is in the detecting devices: an unstable situation of some kind is built up, requiring only a slight irregularity, such as the ionization of an atom, to start a process that eventually leads to something visible.

Although it is little used these days, it is convenient for historical reasons first to take a look at the workings of a *cloud chamber*. The chamber essentially consists of a box containing a gas saturated with alcohol vapour. A piston at one end serves to expand the gas and this brings about an adiabatic cooling. The vapour is now in a supersaturated condition and so needs to condense. When water vapour in the atmosphere finds itself in such an unstable state it forms droplets on dust particles, but in the cloud chamber one carefully excludes the presence of dust. Under such clean conditions one is then able to observe that condensation can also occur on ionized atoms. Once the tiny droplets begin to form they keep growing until they become

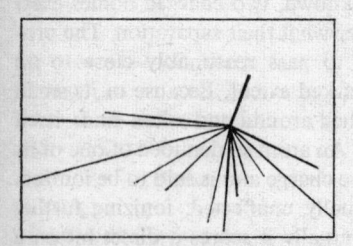

Figure 3 A line drawing of a cloud-chamber photograph showing a nuclear disintegration caused by a cosmic ray

large enough to be seen, whereupon they are photographed by cameras looking through a window in one side of the box. We say cameras in the plural because, in analogy with our own two eyes, one needs to see the trails, or 'tracks', of droplets from more than one vantage point in order to appreciate their three-dimensional layout. The photographs having been taken, the gas is compressed to its former volume and the droplets vaporized.

After a suitable delay, called the 'dead time', to allow the gas to settle down again, the chamber is ready for another expansion.

One serious drawback of the instrument is the tenuous nature of the gaseous medium. The chance of observing a collision on a nucleus belonging to the gas is exceedingly remote. The advent of accelerators underlined its limitations still further, for then bursts of particles became available every few seconds whereas, owing to the dead time of the chamber, a picture could be taken at best only every minute.

Clearly, a detector was needed that employed a naturally dense medium and had a fast recycling time. Such a detector is the *bubble chamber*. In this device a transparent liquid is subjected to a pressure, on the sudden removal of which the liquid expands a little and becomes superheated. (Contrary to what one might think liquids are not incompressible.) In order for boiling to commence, though, the liquid must have some 'centre' from which to start a bubble. The reader is no doubt already aware of this from noticing the way in which bubbles in a beaker of boiling water grow from some irregularities or specks of dust on the walls or bottom of the container. Happily for the physicist, ionization can also provide suitable centres for bubble growth. Thus, in the bubble chamber, the paths of particles show up as trails of bubbles. After being photographed, the trails collapse and disappear under the renewed pressure. Any unwanted liquid motion quickly dies out under the action of the viscous forces and this, combined with the rapidity of the bubble growth, means that the chamber is soon ready to expand again. In fact bubble chambers can keep pace with accelerators.

There are two main types of bubble chamber: one is designed to operate at low temperatures with liquid hydrogen and the other uses liquids incorporating heavy elements. The former allows one to carry out fundamental experiments involving collisions on single protons (for we recall that the nucleus of a hydrogen atom is simply one proton), but liquid hydrogen is still a rather low-density medium, and for certain experiments it is better to use a heavy-liquid chamber where the chances of a collision are so much better.

Modern bubble chambers are big and complex. For instance, the University College London heavy-liquid chamber, which operates alongside the 7 GeV accelerator at the Rutherford High-Energy Laboratory in Berkshire, weighs 200 tonnes. Moving such a colossus from one beam to another could present quite a problem. However, the difficulty is overcome in a rather novel fashion with this particular chamber: it rises up on a cushion of air, hovercraft style, and can then be easily manoeuvred. As just one example of the taxing demands the construction of these devices is making on industry, one may cite the need for a window through which to view the tracks. In order to withstand the fifty atmospheres pressure exerted on it by the liquid, the University College chamber required a slab of optically perfect glass weighing half a tonne. The time taken simply to anneal this window was three months.

One of the great advantages of the next detector we consider,

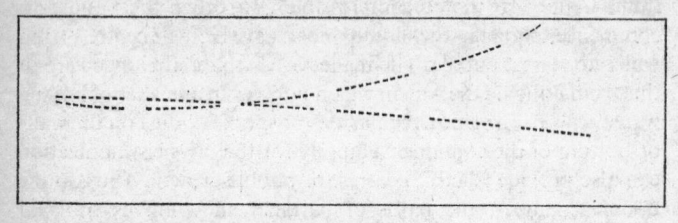

Figure 4 Particle tracks in a spark chamber

the *spark chamber*, is the comparative ease with which it can be made and adapted in shape and size to the requirements of individual experiments. Essentially it consists of a stack of parallel

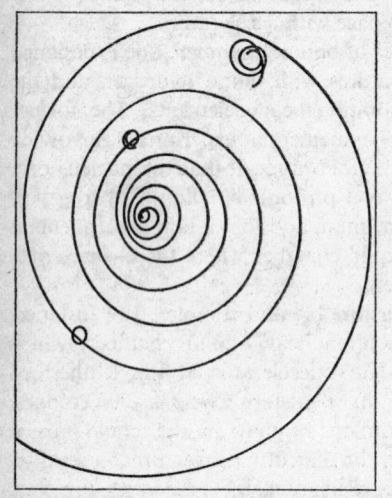

Figure 5 An electron in a magnetic field. The track spirals with decreasing radius of curvature as its momentum is progressively reduced through loss of energy by ionization and radiation

metal plates or wire grids. These are connected up alternately to a high-tension supply, so that a strong electric field is maintained in the gaps between them. The passage of a charged particle causes ionization of the gas between the plates and this leads to a discharge. The positions of the sparks in the successive gaps locate the path taken by the particle through the instrument.

A further method employs *nuclear emulsion*. This is similar to the ordinary photographic emulsion on commercial film, except that it is considerably enriched with silver bromide and thus made more sensitive. It is also manufactured in thick layers rather than in the thin smears one normally finds on celluloid. A silver bromide crystal, besides being liable to activation on exposure to light, can be similarly affected by the passage of a charged particle through it. Consequently, without ever exposing the emulsion to light, particles from the accelerator are fired through it; on development the paths of the particles show up as trails of blackened grains. The grains are very small, so when viewed under a microscope they allow one to study the fine detail of nuclear collisions.

These techniques, together with various types of *counters*, constitute the armoury of the high-energy physicist. In the next article we shall take a look at what he has so far been able to discover with them.

7 F. R. Stannard

High-Energy Nuclear Physics II. Fundamental Particles

F. R. Stannard, 'High energy nuclear physics, II. Fundamental particles' *Physics Education*, vol. 1, 1966, no. 4, pp. 228–35.

The edifice of science not only requires material, but also a plan. Without the material, the plan alone is but a castle in the air – a mere possibility: whilst the material without a plan is but useless matter.

Mendeléef

Methods of measuring mass and velocity

Before proceeding to describe the properties of the fundamental particles, it is perhaps appropriate first of all to say a word about the methods of assessing mass and velocity. What characteristics does one look for in a bubble-chamber or nuclear-emulsion track that enable one to identify the particle?

In a bubble chamber one important aid to identification is provided by a strong magnetic field acting across the volume of the liquid. This serves to deflect the trajectory of each particle into an arc of a circle, the radius of curvature of which is proportional to the particle's momentum (i.e. *mass × velocity*). Whether the deflection is to the left or right depends upon the sign of the particle's charge.

Superimposed upon this constant curvature is a random or spurious wandering from side to side. This is due to the fact that every time the particle ejects an electron from an atom, it is slightly deviated itself by the electric field of the nucleus and its surrounding electrons. There is as much chance of a deflection one way as another, but owing to statistical fluctuations one obtains detectable deviations. The lighter and slower the particle the greater the effect, so measurement of this form of scattering gives information on a quantity dependent upon *mass and velocity*.

The number of ions created per unit length of track depends

upon the time the particle's electrostatic force is able to act strongly upon the electrons in the atoms. Thus the number of bubbles or emulsion grains per unit length is a function of the particle's *velocity*.

Furthermore, the particle loses energy through ionization and so it is possible for all its kinetic energy to be dissipated before it leaves the detecting medium. Such particles as come to rest allow one to assess their original kinetic energy from a simple measurement of the total length of their tracks.

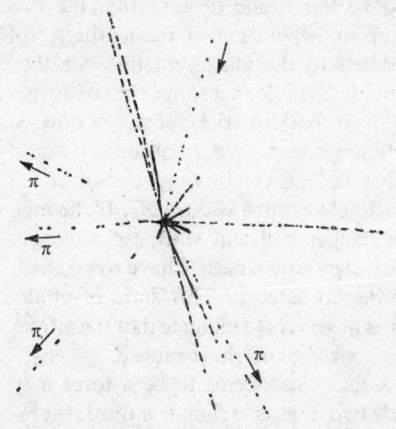

Figure 1 An interaction in nuclear emulsion caused by a cosmic ray particle entering from the top. Several pions are produced

Measurements like these in effect present one with a set of simultaneous equations from which the unknowns, mass and velocity, can be evaluated.

Exchange forces

The interesting idea that there might be elementary particles in addition to the electrons, protons and neutrons was first suggested in 1935 by Yukawa, a Japanese physicist. The problem he faced concerned the stability of nuclei. These were known to consist of nucleons, that is both neutrons and protons, the latter having positive electric charge. Because like charges repel, there must be a strong repulsive force trying to split nuclei apart. Clearly, in order to hold the nucleons together some other force must also exist; one that is attractive and sufficiently strong to be able to swamp the electrostatic repulsion. This 'nuclear force' must act only over a very small distance since it seems not to have any detectable effects on a macroscopic scale. Yukawa pictured it as being due to what one calls an 'exchange' mechanism. The meaning of this term can be illustrated in the following way. Let us suppose there are two cricketers in a distant field throwing and catching a ball. If they are so far off that we are unable to see the ball passing between them, what are we likely to conclude about their behaviour? We would observe that the two men remained within a certain separation of each other. Not knowing that this was related to the maximum distance they could throw the ball, we might conclude that some kind of attractive force existed between them. A third cricketer might now go over and join in the catching practice. We as observers would interpret this by saying that he had come within range of the force and as a result was attracted to the two others. If the men were now to abandon the cricket ball and start exchanging a shot, as used in shot-putting, they would clearly have to be much closer to each other in order to catch it. This state of affairs would no doubt seem to us as observers to indicate that the attractive force was now extremely strong and short-ranged.

The point to be made is that what seems to be a force may simply be a process whereby two objects exchange a third. Moreover, it can be seen that the range of the supposed force is related to the mass of the third object. Forces capable of such a description are known as 'exchange forces', and this is the concept that Yukawa applied to the attraction between nucleons. He was able to predict from the known separation of nucleons in a nucleus, viz. about 10^{-13} cm, that the mass of the exchanged object would be roughly two hundred times that of an electron. As this mass would be intermediate between that of the electron and the nucleons, he proposed calling it a *meson*.

The production of a visible meson in a nuclear collision can crudely be regarded as similar to knocking one of our cricketers sideways just as he is about to catch the ball and thereby making him drop it. Of course, one condition that must be satisfied before a meson can be produced is that there must be sufficient kinetic energy involved in the collision to provide the meson's mass.

Mesons and hyperons

In 1947, twenty-two years after Yukawa's theory was first put forward, it was triumphantly confirmed by the discovery of the predicted particle. It was found to have a mass of 273 electron masses (m_e) and is now called the π-meson or pion. A study of the properties of this particle, it was felt, would soon lead to a good understanding of nuclear forces.

However, the period of self-congratulation was short-lived, for in the same year, quite unexpectedly, additional particles suddenly came to light. There were some called K-mesons which had masses roughly one-half of that of a proton. There was the lambda (Λ) particle with a mass of 2183 m_e which, being heavier than the proton, could not really be called a meson. It was designated a *hyperon* along with heavier ones called the sigma (Σ) and xi (Ξ) which were discovered a little later. Moreover, it soon became clear that some mysterious rules seemed to be operative in the production of these extra particles. It was not sufficient to satisfy the laws of conservation of charge, momentum and mass-plus-energy. These new particles were never created singly; they always appeared from nuclear collisions simultaneously with others.

The mystery deepened still further when one came to consider the subsequent fate of mesons and hyperons. Yukawa had already pointed out that the pion might well be an unstable object. This idea stemmed from the knowledge that a neutron in free space was unstable, and because of its slightly greater mass 'decayed' into a proton, electron and neutrino:

$$n \rightarrow p + e^- + \bar{\nu}_e.$$

The neutrino is an object having no mass but capable of carrying kinetic energy and momentum. The fact that a massless particle can do this is but one of the strange consequences of Einstein's theory of relativity. Yukawa thought that the decay of the neutron could somehow be attributed to an instability in the pion, and thus he thought the pion would decay to an electron and neutrino. However, it was found that the vast majority of pions did not decay directly to an electron but changed to another completely unexpected particle of mass 207 m_e called the muon (μ). The muon is in many respects the most puzzling of all the new particles for it is not created in nuclear collisions; whenever it appears it is as a result of a decay process. This particle itself decays into an electron and two neutrinos.

Like the pion, all the other new particles quickly decay. The K-mesons, for example, can change into two or three pions which themselves subsequently decay. Alternatively, they can miss out the pion intermediate state and decay directly to a muon and neutrino, or indeed to various combinations of the pion, muon, electron and neutrino. The end-product of all these various

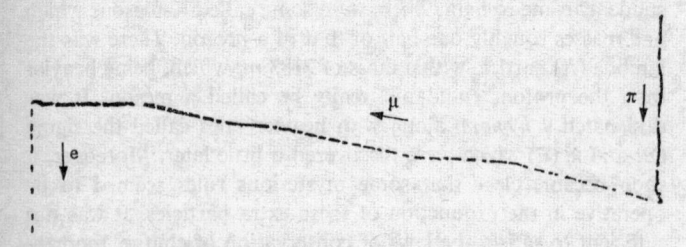

Figure 2 A nuclear emulsion event in which a pion decays to a muon, which in turn decays to an electron. Note the way the ionization increases along the muon's track showing that it is slowing down. (Powell, C. F., 'Mesons', *Report on Progress in Physics*, vol. 13, 1950, pp. 352–424 (London: Physical Society))

chain processes though is the same: a handful of electrons and neutrinos. The original mass of the K-meson almost entirely transforms back into energy.

The Λ-hyperon decays as follows:

$$\Lambda^0 \to p + \pi^-.$$

The lambda incidentally has no charge so cannot leave a trail of ionized atoms behind. We must mention briefly therefore how it is possible to determine the properties of something that is invisible. All the information one needs is to be found in an analysis of its decay. The secondary proton and pion, being charged, leave tracks that can be analysed in the usual ways. From these measurements the total energy and momentum of the particles can be found, and by the conservation laws these must equal the total energy and momentum of the original particle. It is an easy matter then to determine the mass and velocity of the unseen particle. Moreover, the direction of the resultant momentum of the proton and pion must coincide with that of the parent particle, so one may then look along this direction to see the particular nuclear collision which produced the lambda. Knowing the distance the lambda travelled before it decayed, and also its velocity, one may now determine how long it lived.

Returning to the decay scheme of the lambda, one might think

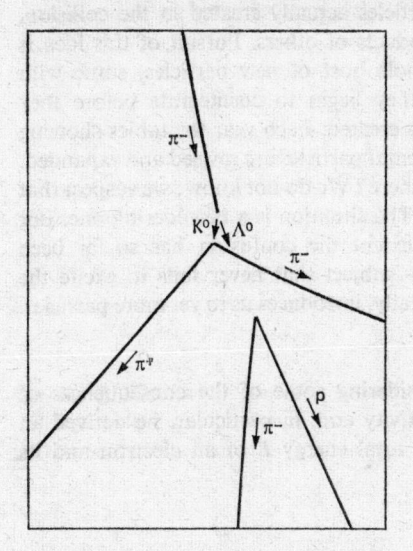

Figure 3 A negatively charged pion interacts in a bubble chamber and produces simultaneously a Λ° hyperon and a K° meson. Neither of these particles leaves a visible track because it carries no charge. They can be identified only when they decay, the former to $(p+\pi)$ and the latter to $(\pi^+ + \pi^-)$

that because the secondary proton is stable, one had this time managed to produce a sizable amount of new stable matter. This however is not so, for in the production of the lambda one originally had to lose a proton; the decay merely returns it. This is the case with all hyperons. At their production a nucleon disappears only to be found again among the eventual decay products.

The lifetimes of most of the new particles lie in the range 10^{-10}–10^{-6} s. These times seem very small until one recalls that the particles are travelling with speeds in the region of 10^{10} cm s^{-1}. They therefore travel distances of the order of a centimetre or more in the detecting medium so their tracks are easily seen.

It was only in the last few years that one really began to appreciate that there could exist particles that were so short-lived that they decayed before they were able to leave a visible track in the detector. The particles we observe coming from a collision may well not be the particles actually created in the collision, but rather the decay products of others. Pursuit of this idea is currently revealing a whole host of new particles, some with lifetimes so short that they begin to disintegrate before they have really finished being created. Each year the tables showing the properties of fundamental particles are revised and expanded. How many particles are there? We do not know; we suspect that there is no end to them. The situation is a bewildering one. But before we see to what extent the confusion has so far been resolved let us turn to a subject that never fails to excite the imagination and, incidentally, introduces us to yet more particles.

Antimatter

In 1928 Dirac was considering some of the consequences of Einstein's theory of relativity and, in particular, he derived an expression involving the total energy E of an electron and its momentum p:

$$E^2 = p^2c^2 + m_e^2c^4,$$

where c is the velocity of light. To solve for E one has to take the square root of both sides and, as is usual when taking the square root, there appears on the right-hand side a $\pm$ sign. The negative solution leads to what on the face of it is an unphysical situation in which there are particles with negative mass that move in the opposite direction to any applied force. In order to make such an object come towards one, it has to be pushed away! In the circumstances, one might have been tempted to quietly ignore this negative solution. However, if one does not, and instead follows through the logical implications of the equation, one finds that although it is possible to explain away the non-existence of negative-mass particles, nevertheless it is necessary to postulate the existence of a new form of matter: *antimatter*.

Corresponding to each type of fundamental particle there is an antiparticle. Particle and antiparticle have the same mass and lifetime, but their electric charges are opposite in sign, as are some other intrinsic properties they possess. Thus an antiproton is negatively charged and an antielectron, generally known as a positron, is positively charged.

One striking property of antimatter is that when it encounters ordinary matter there is mutual annihilation. Thus the proton (p) the antiproton ($\bar{p}$), both of which are normally completely stable and indestructible, annihilate when they meet to produce, for example, energetic pions

$$p + \bar{p} \to \pi^+ + \pi^- + \pi^0.$$

The antiparticle of the neutron has no charge and so looks deceptively like a neutron. In fact the way to tell whether an unknown particle is a neutron or antineutron is to bring it up to a neutron and see whether annihilation occurs.

The reader may be thinking that because the energy release in such annihilations is so great, all one has to do is to create some antiprotons and allow them to collide with ordinary matter to have a useful source of nuclear power. Unfortunately such is not the case because it is found that in order to create an antinucleon one must put into the collision sufficient energy to produce simultaneously a new nucleon. The annihilation process simply restores the energy previously expended.

The existence of antimatter gives rise to endless speculation. For instance, how do we know that all parts of the universe are composed of the same type of matter as our own? We do not, and indeed it is quite possible that some galaxies are made of antimatter with atoms consisting of antinucleons and positrons. High-energy physicists in such galaxies probably regard the tiny amounts of matter they make equally as exotic as the antimatter we produce. It is sad to reflect that a spaceman from an antigalaxy, no matter how friendly his intentions, could not land on Earth without his arrival being heralded by an explosion equivalent to 100 000 nuclear bombs the size of that dropped on Hiroshima.

Laws and classification schemes

We have come a long way since Yukawa first propounded his notions on nuclear forces. His picture of the forces as being due to the exchange of pions, although a useful starting-off point, is now seen to be much too naïve. It appears that when the nucleons are very close together they perform a kind of juggling act, exchanging many different particles.

Many questions arise. Why can one produce particles with only certain discrete masses? What determines the values of those masses? Why are some particles created only in conjunction with others? What determines how long a particle will live and which particles it will change into when it decays? Why should the proton be stable? Where does the muon fit into the scheme of things? Are some particles more fundamental than others, i.e. can some be treated as simply particular combinations of others? Are all the particles we regard today as fundamental merely composite structures of some hitherto undiscovered basic building blocks? If a particle starts to disintegrate before it has finished being made, are we even sure any longer that we know what we mean by the word 'particle'? The list is inexhaustible. Truly we must be in a similar position to that in which Mendeléef found himself when confronted by all the elements and their properties. The periodic table was his means of bringing order to apparent anarchy. We are accumulating material of a different kind; where is our plan?

First of all one may state that in all collisions and decays, jointly known as 'interactions', certain conservation laws must be rigorously obeyed. There are the conservation laws of: *electric charge*, *mass-plus-energy*, *momentum*, *angular momentum*. The first three have already been mentioned. The last is also necessary since each type of particle can spin like a top with a certain angular momentum. For example, the nucleons Λ, Σ, Ξ, μ, e and ν all have the same spin whilst the π- and K-mesons have no spin. In any interaction the sum of the spins of the original particles, together with their orbital angular momentum with respect to each other, must be passed on to the final particles.

Next, one must incorporate into the laws the fact that when a hyperon is produced a nucleon disappears, and when it decays its place is taken once again by a nucleon. It is convenient to give the nucleons and hyperons a collective name, the *baryons*, and then postulate that the number of baryons must not be altered by any interaction. But, one may ask, what about the annihilation of nucleons by antinucleons? To overcome this apparent anomaly one adopts precisely the same procedure as for the case of electric charge. Two charged particles can interact and produce two uncharged particles and yet one still preserves the conservation law of charge. This is done simply by requiring one of the particles to have one unit of positive charge, the other an equal amount of negative charge. It appears justified therefore to keep the law of baryon conservation intact by assigning one positive unit of baryon number to the nucleons and hyperons, and one negative unit to their antiparticles. Pions and K-mesons can be produced in any numbers, not necessarily in particle–antiparticle pairs, so we assign baryon number zero to them. The concept of baryon number is closely linked to the stability of the proton. This particle, being the lightest member of the family, cannot decay, for there are no potential decay products that can take its unit of baryon number.

The reader may by now be feeling distinctly uneasy. This is quite natural for whereas one imagines one knows what energy, mass, charge and momentum are 'like', it is impossible to visualize a baryon number. It has no counterpart in the macroscopic world and yet one is forced to accept that, to a fundamental particle, baryon number is a quantity as real and as meaningful as, say, electric charge. Thus to the four previous conservation laws is added a fifth, that of *baryon number*.

Like the baryons, the electron also belongs to a family. This is a rather exclusive family consisting only of the electron and the neutrino. Whenever one disappears the other takes its place so as to preserve the total number of family members in existence. This rule is obeyed except that an electron can be annihilated by a positron. In analogy to the baryon family therefore, one assigns an electron number of $+1$ to the e^- and ν_e, and -1 to the e^+ and antineutrino ($\bar{\nu}_e$). There exists an exactly similar family for the muon and its neutrino (ν_μ). So we may add two further conservation laws to our list, those of: *electron number, muon number*.

Are there any more family resemblances and corresponding

conservation laws between particles? The answer is yes and no. There are certainly other strong affinities between particles as we shall see a little later, but their corresponding conservation laws, unlike those so far enunciated, are not absolutely obeyed; in some processes they are flouted. It is clear that at this point we must say a word about the different types of interaction, so that we shall be able to recognize those situations where a given law is applicable.

The physicist recognizes four types of interaction: *gravitational*, *strong*, *electromagnetic* and *weak*. Although the fundamental particles exert a gravitational attraction on each other, the effects of it are so minute that it will not be considered further.

The *strong* interaction is what we have so far called the nuclear force. It binds nucleons in a nucleus, and it is by this interaction that particles generally have their effect on each other when they collide. The interaction is called 'strong' because it takes place very rapidly. If the incident particle is travelling at about 10^{10} cm s^{-1} and each particle has a radius of about 10^{-13} cm, then they will be in range of each other for only about 10^{-23}–10^{-22} s. A transformation occasioned by a strong interaction must therefore take place in this order of time.

The *electromagnetic* interactions are to be identified with the forces between electric charges, and in particular give rise to the phenomenon of ionization. It is not as strong as the previous interaction.

As its name implies the *weak* interaction, which is responsible for the decay of particles, takes a long time to produce any effect. In this context it is important to realize that a decay time of 10^{-8} s is a 'very long time'. It is not so by normal everyday standards of course, but on the nuclear scale, where the natural unit of time is 10^{-23} s, decays are so slow that they can almost be said not to happen. That this is the case can be seen if we imagine all processes slowed down so that a strong interaction needs 1 second to occur. A weak interaction would then take typically a hundred million years to complete.

As we said previously, these interactions behave differently with regard to the two conservation laws to be described next. The first of these is connected with the observation that some particles are never produced singly in collisions between nucleons.

They always appear in twos and threes. This is interpreted as an indication of the existence of a new number to be assigned to the particles in addition to those already specified. It has been given the colourful name of strangeness number, and its possible values are $S = +1, 0, -1, -2, -3$. The nucleons and pions have $S = 0$, the K^+ $S = +1$, the K^-, Δ and Σs $S = -1$, the Ξs $S = -2$, etc. Thus in order to produce a Ξ^- in a pion–

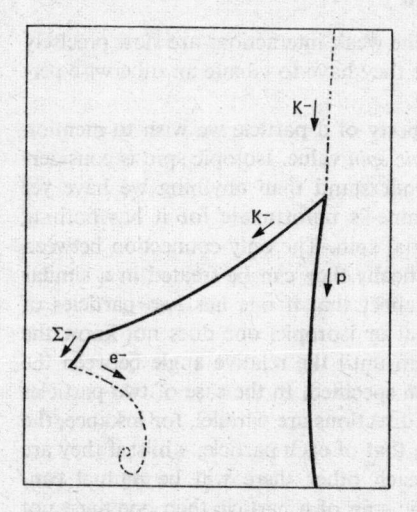

Figure 4 A K^- meson enters a bubble chamber and knocks a proton out of a nucleus. It re-emerges and makes a second collision, this time passing its unit of strangeness on to a Σ^- hyperon. The Σ^- decays into a neutron and neutrino, which are unseen, and a negative electron. (Line drawing of a photograph taken in the Ecole Polytechnique heavy-liquid bubble chamber)

neutron collision it is necessary to create simultaneously two $S = +1$ particles, for example

$$\pi^+ + n \rightarrow \Xi^- + K^+ + K^+$$

Strangeness $0 + 0 = -2 \;\; +1 \;\; +1$
Charge $1 + 0 = -1 \;\; +1 \;\; +1$
Baryon No. $0 + 1 = \;\;\;1 \;\; +0 \;\; +0$

For the above reaction we show in detail how the strangeness, charge and baryon number remain unaltered.

F. R. Stannard

The strangeness conservation law is obeyed by both the strong and electromagnetic interactions, but does not apply to weak interactions such as the following decay

$$\Sigma^+ \to p + \pi^0$$

Strangeness $-1 \neq 0+0$
Charge $1 = 1+0$
Baryon No. $1 = 1+0$

In fact one can say that the weak interactions are slow precisely because in order to occur they have to violate an otherwise perfectly good law of nature.

There is one final property of a particle we wish to mention here and that is its *isotopic spin* value. Isotopic spin is considerably more difficult to understand than anything we have yet encountered. Even its name is unfortunate for it has nothing to do with ordinary spatial spin. The only connection between the two is that mathematically they can be treated in a similar fashion. This means in effect that if one has two particles of known spin, either spatial or isotopic, one does not know the overall spin of the system until the relative angle between the two components has been specified. In the case of two particles with equal spins, if their directions are parallel, for instance, the overall spin will be twice that of each particle, whilst if they are aligned antiparallel to each other there will be mutual cancellation. For the isotopic spin of a particle then one must not only accord a value but also a direction in some 'isotopic spin space'. Once again we find ourselves forced into accepting an unfamiliar concept. It is nevertheless vitally necessary, for in the strong interactions the isotopic spin of a system is found to remain unaltered. Electromagnetic and weak interactions do not obey this final law which we now add to our list: conservation of *isotopic spin*.

Thus we see that an important task of the physicist is to assign to each particle appropriate values for its charge, isotopic spin, strangeness number, baryon number, etc. Having done so he is well on the way to predicting how it will behave in any given set of circumstances.

In the last few years there has been a great new advance in methods of particle classification. As a result of this, one is now able to discern that many particles, hitherto thought to be unrelated, do in fact belong to a larger group of particles with similar characteristics. These groupings contain 1, 8 or 10 particles. Furthermore, one has now learnt how to link these groupings to each other to form even larger sets of 1, 35 and 56 particles. The numbers of particles appearing in the groupings at first

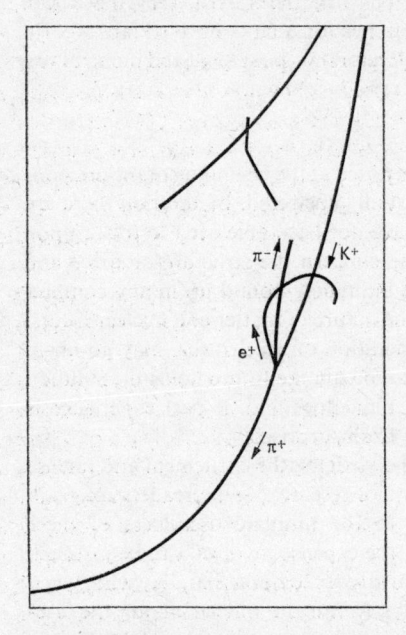

Figure 5 A K^+ meson decays in a bubble chamber by the following scheme: $K^+ \rightarrow \pi^+ + \pi^- + \nu_e$. Only 1 in 30 000 K^+ mesons decay according to this mode. (Line drawing of a photograph taken in the W. Powell heavy-liquid bubble chamber at The Lawrence Radiation Laboratory, Berkeley)

sight might seem rather arbitrary. There is in existence however a theory which requires just these characteristic numbers, a theory describing systems built up in a special manner from three basic components. (A demonstration of how this connection arises unfortunately demands a considerable knowledge of advanced mathematics and so cannot be attempted here.) What are the three basic components? It is quite possible that they are merely convenient mathematical quantities bearing no direct correspondence to any physical objects. On the other hand there may indeed be three truly fundamental particles, which when tightly bound together in certain combinations give rise to all the objects we have to date regarded as the elementary particles. In anticipation that such basic particles exist, they have been called *quarks*. Intensive searches for quarks have so far revealed nothing; perhaps the accelerators we possess at the moment are not sufficiently powerful to produce them.

Concluding remarks

It is clear that the questions posed at the beginning of the previous section have only been partially answered. In addition there are other deep problems we have not been able even to touch upon in these two articles. These concern the structure of space and time, a fascinating subject intimately bound up in any comprehensive understanding of the nature of matter and nuclear forces. The advent of the new generation of accelerators may go far in clarifying the picture. Undoubtedly the future holds the solution to our present tantalizing difficulties and, if past experience is anything to go by, many a fresh surprise.

And yet, while fully acknowledging the excitement and intense interest of high-energy physics research, some readers may still have lingering doubts as to the ultimate usefulness of these investigations and whether the expense involved can be justified. One is tempted by way of an answer to quote Faraday who, when questioned by a sceptical government official about the usefulness of his experiments on the magnetic effects of electric currents replied, 'I do not know, but I am sure that some day you will be able to collect taxes from its applications'.

Alternatively one may adopt the stand of Professor V. Weisskopf, the retiring Director General of CERN, and it is with his words that we conclude:

The pursuit of fundamental questions was and is the spearhead of science. It attracts the most sophisticated brains and it supplies vitality and vigour to the scientific community which benefits the totality of scientific development. If the spearhead is blunted the physical sciences as a whole would suffer.

8 C. Ramm

High-Energy Neutrinos

Colin Ramm, 'High energy neutrinos', *Science Journal*, vol. 4, 1968, no. 4, pp. 56–63.

Neutrinos are the commonest particles in nature. There are more of them than there are atoms in the universe and their total energy is much more than that of all the visible stars. Why then did Wolfgang Pauli, who first postulated their existence, wager that neutrinos would never be detected? The answer, of course, is that neutrinos interact with matter so incredibly rarely. The first experiments designed to look for them, years before Pauli's postulation, failed in sensitivity by a factor of 10^{20}. Low-energy neutrinos can cross the whole universe with very little chance of interacting; there is no place which is shielded from the penetration of these neutral and massless particles.

Neutrinos coming from the sun bombard the Earth's surface at a rate of 10^{14} neutrinos per square metre per second. Their energies are typically one million electronvolts (1 MeV). These 'low'-energy neutrinos carry several per cent of the sun's energy; if they interacted with our surroundings as readily as photons, which are also neutral and massless, they would make our Earth uninhabitable. Instead, they transmute only a few atoms per year in every tonne of matter. No experiment has yet been sufficiently sensitive to detect their effects.

Thus, compared with any other particles, such neutrinos are virtually aloof from interactions with matter. They have no known interactions with electric or magnetic fields and are confined within the boundaries of space only by those aspects which make energy and mass equivalent manifestations of matter. It is not surprising then that the force that governs the interactions of neutrinos is known as the 'weak' force.

High-energy neutrinos are somewhat less elusive. Those of energies of a few GeV (a few thousand million electronvolts) –

typical energies in present 'high'-energy experiments – interact with protons and neutrons about a million times more readily than low-energy neutrinos but their cross-section for interaction is still only about 10^{-38} cm^2. Cross-section is the parameter which measures the probability of interaction; it is the apparent surface which the two interacting particles present to each other. High-energy neutrinos could travel in lead for a distance equal to that which separates the Earth and the sun before acquiring much probability of interacting with an atom of lead. Nevertheless, despite Pauli's wager, the very rare interactions of such particles do occur and can be observed and studied in detail in the laboratory. More than a thousand interactions of high-energy neutrinos, in a cubic metre of liquid lighter than water, were photographed at CERN, Geneva, late last year. In this article I shall discuss such experiments at CERN and at other laboratories that are opening up a vast new field of physics, which is among the most fascinating and most challenging of modern science.

In the sun and other stars neutrinos (ν) are produced in those nuclear processes by which light elements are converted to heavier ones. For example, the fusion of two protons to form deuterium produces a positron and a neutrino. Nuclear reactors, which essentially transform heavier elements into lighter ones – a process akin to radioactive decay – lose considerable energy in the form of antineutrinos ($\bar{\nu}$). When a neutron decays into a proton, an electron and an antineutrino are produced. The ν and $\bar{\nu}$ from nuclear processes have energies up to a few MeV. Most of the unstable elementary particles also have some mode of decay which produces neutrinos; pions and kaons, even at rest, produce more energetic neutrinos than any nuclear transformation.

Various processes of neutrino production are shown in Figure 1(a). The neutrinos produced are not all the same. Some processes yield the antiparticle of the neutrino – the antineutrino. There are also two distinct families depending on whether the neutrino is produced in association with an electron or a muon. This fact, which was long suspected because the muon decays only into an electron and two neutrinos, was confirmed in the first high-energy neutrino experiment, which I shall discuss later in the article. Unless it is necessary to distinguish between them, they are all called neutrinos, whether they are ν_e, the electron neutrino; $\bar{\nu}_e$, the electron antineutrino, the one Pauli postulated; ν_μ, the muon neutrino; or $\bar{\nu}_\mu$, the muon antineutrino.

in stars	neutron decay	pion decay	kaon decay
$p^+ + p^+ \rightarrow d^+ + e^+ +$ $+ \nu_e + 0.4$ MeV	$n \rightarrow p^+ + e^- + \bar{\nu}_e +$ $+ 0.8$ MeV	$\pi^+ \rightarrow \mu^+ + \nu_\mu +$ $+ 34$ MeV	$K^+ \rightarrow \mu^+ + \nu_\mu +$ $+ 38.8$ MeV
(a)		$\pi^- \rightarrow \mu^- + \bar{\nu}_\mu +$ $+ 34$ MeV	$K^- \rightarrow \mu^- + \bar{\nu}_\mu +$ $+ 38.8$ MeV

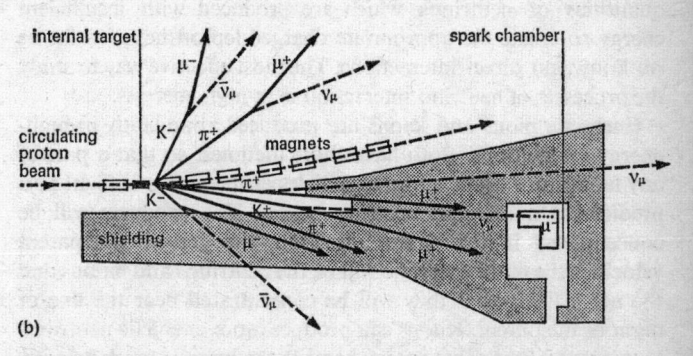

(b)

Figure 1 Neutrino production takes place by way of the mechanisms shown in (a). In stars two protons fuse to form deuterium together with a neutrino and a positron. Neutron decay results in the production of an antineutrino. Both these processes yield 'low-energy' particles. Pion and kaon decay, however, both yield higher-energy particles and it is these two 'parents' which are employed in neutrino detection experiments with high-energy accelerators. In the first high-energy neutrino experiment at Brookhaven, shown in (b), a proton beam strikes a target placed inside the 31 GeV synchrotron and produces a stream of secondary pions and kaons. These in turn decay to a mixed beam of neutrinos and antineutrinos. Their very rare interactions can be observed in a spark chamber. All other particles are absorbed in the shielding

Neutrinos, electrons and muons are called leptons. Neutrino interactions are always consistent with an empirically observed conservation law which demands that the total number of leptons must not change. The disappearance of a neutrino in an interaction is accompanied by the manifestation of the appropriate charged lepton. For example, a $\bar{\nu}_e$ of sufficient energy can interact with a proton to produce a positron and a neutron, a process which is the inverse of the way in which the $\bar{\nu}_e$ was produced – in fact it is called 'inverse beta decay'. Similarly, a ν_μ of sufficient energy interacting with a neutron can produce a negative muon and a proton. With neutrinos of sufficient energy, other particles may also be produced, but never without the electron or muon. Scattering, in which an incident particle bounces off another and retains its identity – the commonest interaction of all other particles – has never been detected for neutrinos. Thus, the vast quantities of neutrinos which are produced with insufficient energy to create the appropriate charged lepton have, as far as we know, no direct interactions. The most effective way to study the processes of neutrino interaction is at high energies.

Energetic pions and kaons are produced abundantly in high-energy interactions. Both have short lifetimes, so that a part of any laboratory beam of these particles disintegrates in flight into predictable muon and neutrino beams. The products will be concentrated in the parent direction, the greater the parent velocities the more energetic will be the neutrinos and muons and the more likely that they will be concentrated near the line of flight of the parent. Kaons can produce more energetic neutrinos than pions of the same energy because the kaon is much heavier. In its centre-of-mass system the neutrino and muon share almost equally the energy of the rest mass of the parent, whereas in the centre-of-mass system of the pion, which is lighter, the neutrino carries less than one quarter of the energy available. Both pions and kaons produce muons, which are the most prolific neutrino parents of all; the muon decay yields two neutrinos but the muon lifetime is so much longer than pions or kaons that muons are of no practical value for laboratory high-energy neutrino beams. They are, however, an important source of cosmic-ray neutrinos.

All high-energy neutrino beams are produced by directing the greatest possible number of energetic parents at the detection apparatus. Experiments are feasible in those laboratories with an accelerator powerful enough to give intense secondary beams of pions and kaons. Four proton accelerators fulfil this requirement; the 31 GeV synchrotron at Brookhaven and the 12·5 GeV one at Argonne, both in the United States; the 25 GeV

(a) $\nu_\mu + n \rightarrow \mu^- + p$
energy of ν_μ: 1·52 GeV

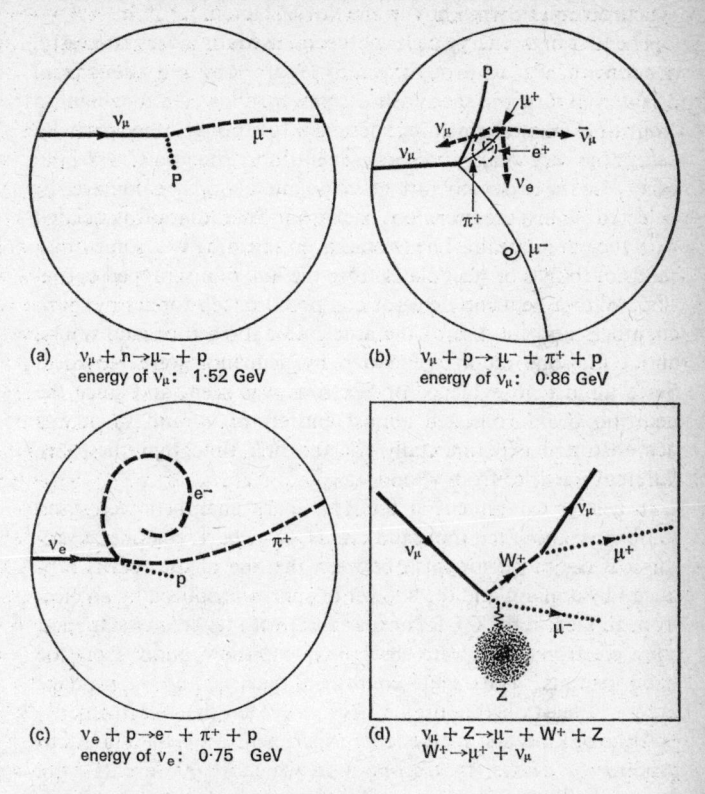

(b) $\nu_\mu + p \rightarrow \mu^- + \pi^+ + p$
energy of ν_μ: 0·86 GeV

(c) $\nu_e + p \rightarrow e^- + \pi^+ + p$
energy of ν_e: 0·75 GeV

(d) $\nu_\mu + Z \rightarrow \mu^- + W^+ + Z$
$W^+ \rightarrow \mu^+ + \nu_\mu$

Figure 2 Neutrino interactions of the type shown under each set of diagrams can be detected in laboratories with accelerators powerful enough to provide intense beams of parent pions and kaons which in turn produce neutrinos. Reconstructions of neutrino interactions photographed in the CERN heavy-liquid bubble chamber are shown in (a), (b) and (c). (d) depicts an interaction which has never been observed: the predicted creation and decay of the intermediate boson, W – the possible carrier of the weak force. Positive bosons could be created from neutrinos, negative ones from antineutrinos. The lifetime of the W is expected to be short – it would decay rapidly into a muon and a neutrino. Since no evidence of the boson has been found it has been deduced that if it exists its mass must exceed 2 GeV: this is higher than first thought

synchrotron at CERN and, since September 1967, the 76 GeV synchrotron at Serpukhov in the Soviet Union.

The first high-energy neutrino beam, used for a very successful experiment, was at Brookhaven in 1962. Pions and kaons from an internal target in the synchrotron were allowed a free flight in front of a massive shielding, designed to stop all other particles except the very weakly interacting neutrinos. Shielding is essential otherwise the extremely rare neutrino interactions could never be detected against the enormous background of radiation associated with the parent beam. The Brookhaven shielding was some thousands of tonnes of steel plates from the hull of a scrapped battleship. With a neutrino detector composed of ten tonnes of spark chambers and full use of the accelerator for a thousand hours, about fifty-one muons generated by neutrinos were recorded. Since almost no evidence of electrons was seen, and since the neutrino beam consisted almost entirely of ν_μ and $\bar{\nu}_\mu$, it was demonstrated experimentally for the first time that these are different particles from ν_e and $\bar{\nu}_e$.

In a later experiment at CERN, spark chambers were used to observe about ten thousand events. In a spark-chamber detector it is easy to distinguish between the line of the sparks produced by a muon and the shower of sparks produced by an electron. In fact, in the CERN experiment with its greater statistics, a few electron events were observed due to the ν_e and $\bar{\nu}_e$ from the kaon parents. The result confirmed that ν_μ and $\bar{\nu}_\mu$ produce muons at least one hundred times more often than electrons.

Although the first high-energy experiment took 1000 hours to produce 51 events, the second – in which 10 000 events were recorded – obviously did not take the equivalently longer time which would have been about 200 000 hours or more than 20 years! This was because the intensity of the neutrino beam had been increased about a hundred times in the CERN experiment. This was done by means of two major advances: the beam was first extracted from the synchrotron so that forward moving protons could be used for the production of the parents for the neutrinos; and the neutrino parents were focused to increase the probability of their traversing the detector.

The neutrino parents from the target where they are produced by the proton beam have a wide range of energy and direction.

There are no known principles by which such a collection of parents can be focused perfectly to a point. Nevertheless, ingenious magnetic devices have been invented which guide many of the parents which are not travelling towards the neutrino detectors into a more useful direction. The first of such inventions was the magnetic 'horn'. During the 2·1 ms for which the proton beam bombards the target, two coaxial conductors carry a current of 300 000 A. If the flow along the inner conductor is in

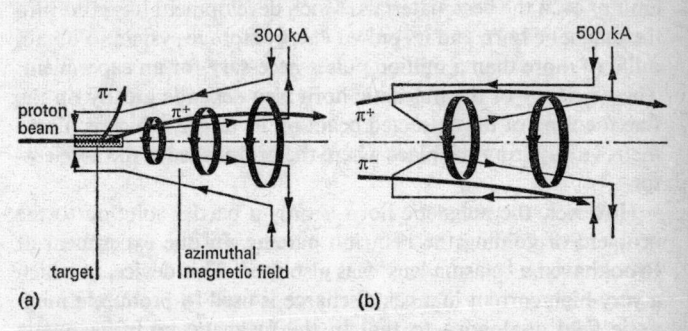

Figure 3 Focusing devices have been constructed to guide parent particles towards the neutrino detector. The magnetic horn, shown in (a), was the earliest of such devices. The proton beam bombards a target for 2·1 ms during which time the two coaxial conical conductors carry a current of 300 000 A. This current produces an azimuthal magnetic field which deflects charged particles towards or away from the axis according to their sign. The principle of the recently-devised magnetic reflectors is shown in (b). Two of these magnetic lenses correct the focusing of neutrino parents emerging from the horn with trajectories steeply inclined to the axis. As with the horn a momentary high current produces an azimuthal field between the two shaped coaxial conductors. The reflectors increase the overall neutrino flux by a factor of three over the horn alone

the direction of the neutrino beam, the azimuthal magnetic field between the conductors deviates positive particles – the neutrino parents – towards the axis of the horn and, of course, negative particles further from the axis. The precise shape of the inner conductor, to obtain the maximum concentration of neutrino parents for a particular proton energy and target, is determined from computer studies. The first magnetic horn increased the total neutrino flux by a factor of six.

Unfortunately, the neutrino parents traverse the actual conductors. Therefore, to reduce losses from unwanted interactions the conductors and the target must be as thin as possible; at its narrowest place the diameter of the inner conductor of the present horn is only 15 mm. Such a conductor can carry a current of 300 000 A only momentarily because the mechanical shock both from the magnetic field and from the sudden heating caused by the current produces stresses of many tonnes, close to the limit of even the best materials. Much development has gone into the magnetic horn and its pulsed energy storage system to obtain a life of more than a million pulses necessary for an experiment. The efficiency of the magnetic horn also depends greatly on the fine focusing of the extracted beam on its target, which is ninety metres away from the place where the protons leave the accelerator.

However, the magnetic horn is only a partial solution to the problem of guiding the neutrino parents. In one experiment at Brookhaven a 'plasma lens' was also used. This device, in which a very-high current in a gas discharge is used to produce a magnetic field analogous to that in the magnetic horn, has many valuable features. Especially important is the almost complete absence of absorbing material in the paths of the neutrino parents. In practice, however, it is difficult to maintain the distribution of the electric current, and hence the magnetic field, exactly as required.

A new advance in parent focusing has come from the invention of two toroidal lenses, called 'reflectors', which correct the focusing of neutrino parents emerging from the horn with trajectories steeply inclined to the axis. They, too, are guided towards the detectors and increase the overall neutrino flux by three times that obtained from the horn alone. The optical principles are similar to those of the magnetic horn. The ingenuity has been in discovering from computer calculations that elements with these shapes and locations can contribute so much to the neutrino flux.

While shielding cannot absorb a significant fraction of neutrinos, it has a major influence on the intensity of the neutrino beam because the intensity on axis at the detector is inversely proportional to its length; the best shielding is as short as possible consistent with the absorption of unwanted particles. The stopping power of a material is almost proportional to the mass traversed; thus the denser the shielding the more intense the neutrino beam at the detector. In high-energy experiments it is the maximum energy of the unwanted muons from the neutrino parents which finally regulates the choice of shielding length. The way in which the six thousand tonnes of steel ingots, lent by the Swiss government, and many thousands of tonnes of concrete have been arranged in the CERN experiments is shown in Figure 4.

These are the essential material features of what is at the moment the world's most intense high-energy neutrino facility; in pulses two seconds apart some 10^{10} neutrinos traverse the detector to produce about five interactions per tonne of material per hour. As with all the other activities of CERN, this installation has grown from international collaboration. The first magnetic horn was developed by a Dutchman; its power supply by a German; the reflectors were designed by a Yugoslav and a Swiss, and their power supplies were made by a Norwegian; the latest version of the magnetic horn was designed by two Englishmen while at Bristol. The group of more than thirty scientists devoting themselves to the study of high-energy neutrinos in this laboratory come from almost as many countries.

The neutrino event rate is proportional to the mass of material available for interaction. Spark chambers have been assembled with as much as thirty tonnes of target material. The data obtained are numerous but the information per event is low. A particular consequence of the intensity of the CERN neutrino facility is that it has made feasible the use of a heavy-liquid bubble chamber of 1·1 m³ capacity as a detector. The one used is the largest in the world and is in a magnetic field of 27 kilogauss.† In the most recent experiment this chamber was filled with liquid propane (C_3H_8) which contains more hydrogen per unit volume than liquid hydrogen; in high-energy interactions the few electron-volts binding the hydrogen atoms to the carbon in the propane molecule is trivial.

Of the 0·5 tonne of propane in the chamber, about 100 kg is 'free' protons. Such a chamber installed in the beam in which

† [27 kilogauss = 2·7 T].

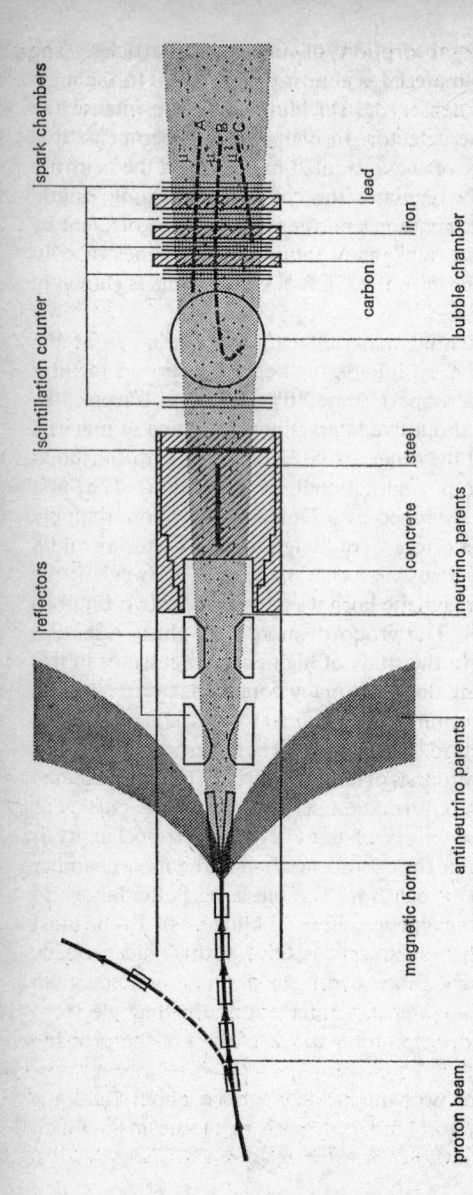

Figure 4 CERN neutrino facility as improved in 1967 uses neutrino parents produced by protons from the 25 GeV proton synchrotron. Unwanted radiation is suppressed by a shielding of six thousand tonnes of steel and thousands of tonnes of concrete. The magnetic horn and two reflectors guide the neutrino parents towards the spark chambers and a heavy-liquid bubble chamber. The diagram shows three types of muon neutrino events: A shows a muon produced at the end of the shielding and which crosses the bubble chamber and the remaining spark chambers; B shows a muon produced in the bubble chamber which also crosses the spark chamber; C is the production of a muon by neutrino interaction in an iron-plate target in the spark chamber. Muon production in the shielding can be identified by the fact that at the moment of the event no charged particle was detected traversing the scintillation counter

the two-neutrino problem was solved in the first experiment would have registered only two neutrino events, each of which would have had about a 20 per cent chance of being on a proton. In the latest operation, which was also allocated about a thousand hours of accelerator time, the proton beam was shared with several independent experiments and the neutrino beam with two spark-chamber experiments. Even under these circumstances about a thousand events were photographed. About 80 per cent of these are in carbon nuclei but when the analysis is completed there will be about a hundred very good candidates for interactions on free protons.

During the experiment a cycle of operation is established so that when the neutrino beam is produced the liquid in the bubble chambers is decompressed. Spontaneous boiling then occurs along the tracks of any ionizing particles in the chamber. Two milliseconds later, when these bubbles have grown to a convenient size, flash lamps are triggered to record the appearance of the chamber in three cameras, the liquid is recompressed to stop the boiling, the films advance and the whole installation prepares for the next operation two seconds later. About 1·1 million pictures were taken in this way on about three hundred kilometres of film. Thus, to obtain one neutrino event required an exposure of about 270 m of film, which contains three views to permit spatial reconstruction of the position of the tracks in the magnetic field in the chamber. The total duration of the neutrino beam for the experiment, which ran between May and November 1967, was only 2·3 s.

While important results have already been obtained from the study of high-energy neutrino interactions, they must be regarded as preliminary; many elementary questions still remain unanswered. Are neutrinos from kaon parents different from neutrinos from pion parents? There is certainly no reason to believe that they are but, on the other hand, there is no significant experimental evidence to show that they are not. Do muon neutrinos ever interact to produce electrons? From the Brookhaven and CERN studies, not as often as once per hundred interactions; but it is of the profoundest importance to know whether the answer is really never or just sometimes. Then we would know if there is some deep connection between v_μ and v_e and perhaps between muons and electrons. To what extent electron neutrinos ever produce muons has not been touched on by present experiments.

In the latest experiment at CERN two simple questions are being investigated further. Does the ν_μ ever produce a μ^+ instead of a μ^-? From the previous work with a spark chamber and a bubble chamber used separately, the answer was not as often as once per hundred times; on this occasion the bubble chamber and a spark chamber have been used together to see whether a μ^+ might be produced as often as once per thousand times. A part of the neutrino shielding was defined as a neutrino target and hence as a source of muons. The positions of emerging particles were indicated in a spark chamber. Some of the particles then went on to traverse the bubble chamber and emerged into another spark-chamber array with dense plates, where their penetrating power could be determined. Thus, muons could be distinguished reliably from pions because the pions would have interacted in the plates. The shielding was designed with great care so that at most only a few μ^+ from the neutrino parents would have pierced it during the whole experiment. An absorber was fitted in front of the target in the magnetic horn to eliminate those antineutrino parents which otherwise would not have been rejected by the parent-focusing magnetic fields.

The three possible origins of any positive muons identified in the bubble chamber could have been: the minute residual penetration of muons through the shielding, the interaction of the small $\bar{\nu}_\mu$ flux, or ν_μ interactions which produced μ^+. It will take many months to determine the relative significance of these possibilities in the data obtained and to establish a new statement on how often a ν_μ can produce a μ^+.

The other question which has been studied in the same installation is whether neutrinos really interact at a rate proportional to the mass of the target nucleus. Interactions between particles which are governed by the 'strong' or 'electromagnetic' forces occur at a rate which is not in general proportional to the mass of the target nucleus. The result for neutrinos will be obtained from a determination of the interaction rate per unit mass in plates of carbon, iron and lead in the spark-chamber array.

Many interesting deductions were made from the data from

the bubble chamber in the 1963–4 CERN experiments. To obtain an adequate event rate with the neutrino facility as it was then, the chamber was filled with freon, CF_3Br, which has a specific gravity of 1·5. A quantitative analysis could be made of the interaction of a neutrino with a neutron, which produces a muon and a proton.

Although the target neutrons were bound in the nuclei of the atoms of freon an analysis was possible because both the muon and the proton had a high probability of escaping from the nucleus where they were formed without further interactions which could invalidate the calculations. How does the ν_μ change to a μ^- and the neutron to a proton?

Many high-energy phenomena can be explained by the concept that all particles are composite states of other particles. A neutron, for example, can be considered as a composite state of a proton and negative particles, or a neutron and neutral particles, although the spontaneous dissociation of the neutron at rest into these particles can never be seen unless it can obey the laws of conservation of energy and momentum. Nevertheless, high-energy scattering experiments can be interpreted as if the neutron is surrounded by a cloud of these virtual particles which are being emitted and absorbed continuously. Some aspects of proton–neutron scattering, for example, appear as a close approach of a proton and neutron in which a neutral pion from the virtual cloud of particles around the neutron is exchanged with the proton. By this means energy and momentum are transferred.

From the statistics of the angular distribution of particles in any interaction it is always possible to define some of the properties of the virtual particles involved in the interaction; sometimes these virtual particles correspond with real particles, sometimes the result may have only a computational significance. In the 1963–4 experiments computations from the angular distribution of the muon and the proton, in about sixty good examples of the interaction, could be interpreted as the transfer from the neutron to the neutrino of a negative particle, of mass 0·6 GeV, from the virtual cloud around the neutron. The data were insufficient to infer in what way this virtual particle corresponds with any other known particles. Another interpretation of the data could be that the neutrino does not interact with the neutron at a point but 'sees' the electromagnetic properties and mass of the neutron distributed over a certain region. Further steps in the elucidation of these fundamental processes will come when the new results are available both from a spark-chamber experiment which was performed at Argonne during 1967 and also from the most recent CERN experiments.

The simplest interaction of a neutrino with a proton is

$$\nu_\mu + p \to \mu^- + \pi^+ + p.$$

When this process takes place on a proton bound in a nucleus, the muon and the proton are likely to be useful for analysis but the pion can be absorbed in the nucleus where it is produced. Thus, the identification and analysis of the event will be uncertain. This fact, together with clear evidence that pion production by neutrinos was more frequent than the elastic interaction, pointed directly to the need for a study in which the target nuclei were free protons. This is one of the experiments in the CERN bubble chamber which has just finished its operational stage. It will still take some months to complete the analysis but it is already clear that many events were recorded which are neutrino interactions on free protons.

Neutrinos differ from all other particles in that they have never been observed to scatter with a transfer of momentum only. As I mentioned earlier, the only known interactions involve a transfer of electric charge as well as a transfer of energy and momentum. Every interaction we have observed is consistent with this statement; many more types of interactions than I have described have been seen but the statistics for each type are so few that it will be necessary to wait for higher total event rates to study them in detail.

In 1965 a small amount of CERN accelerator time was devoted to an experiment with antineutrinos. Their parents are the negative pions and kaons, which are focused by the horn and reflectors when the current in the inner conductors flows against the direction of the neutrino beam. The data obtained were few but it was easy to distinguish interactions of the type:

$$\bar{\nu}_\mu + p \to \mu^+ + n,$$
$$\bar{\nu}_\mu + p \to \mu^+ + \pi^- + p,$$
$$\bar{\nu}_\mu + n \to \mu^+ + \pi^- + n.$$

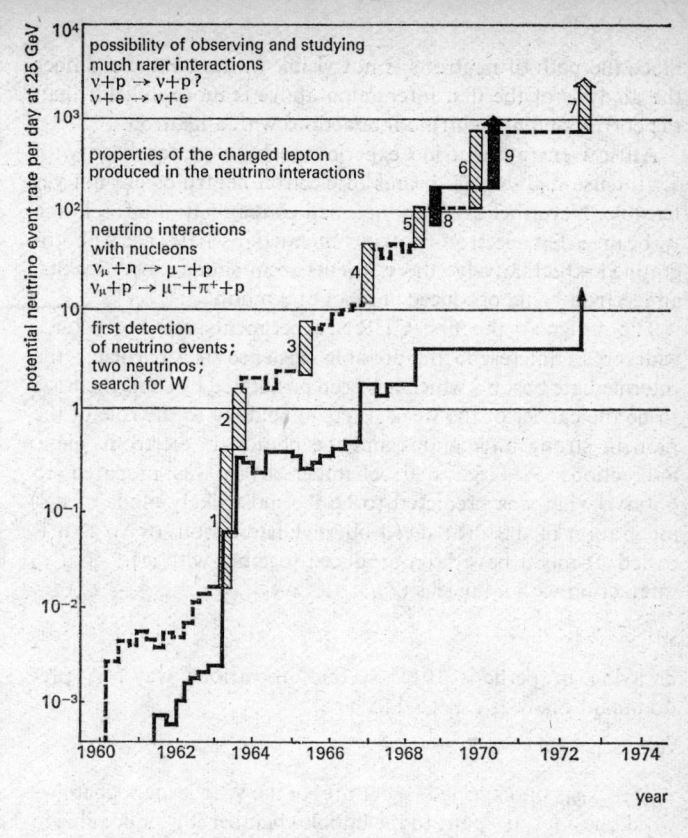

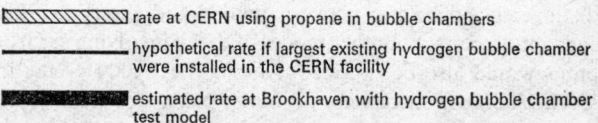

Figure 5 Potential event rate in neutrino facilities at CERN and Brookhaven are shown right for currently planned constructional programmes. Factors resulting in progressive improvements at CERN are: (1) extraction of the accelerator beam; (2) use of the magnetic horn; (3) enlargement of the bubble chamber to 1·1 m^3; (4) construction of the 1967 facility; (5) new magnet power supply; (6) use of the Gargamelle – the 10 m^3 bubble chamber; and (7) estimated contribution from the new injection system. Points (8) and (9) are an estimated situation corresponding to (5) and (7) for the use of the model for the large hydrogen bubble chamber at the Brookhaven facility

Since the path of neutrons is not visible in the bubble chamber the analysis of the first interaction above is more difficult than the corresponding neutrino interactions with a neutron.

All low-energy neutrino experiments have so far been with $\bar{\nu}_e$. Intense high-energy beams of electron neutrinos are not yet feasible. Nevertheless, from the small contamination of ν_e in the ν_μ beam a few electron-neutrino interactions have been photographed. Qualitatively, the ν_e events seem similar to ν_μ events, an electron being produced instead of a muon.

The design of the first CERN experiments was based on a widespread interest in the possible existence of a particle – the intermediate boson – which has been postulated by some theorists to be the carrier of the weak force, in analogy to the role of the pion in strong interactions and the photon in electromagnetic interactions. A large spark-chamber array was prepared to observe what was predicted to be the most likely mode of disintegration of the short-lived intermediate boson, or W as it is called. It might have been produced together with a μ^- in a ν_μ interaction with a nucleus (Z),

$$\nu_\mu + Z \rightarrow \mu^- + W^+ + Z,$$

decaying in perhaps 10^{-18} second in various ways. A predominant one was expected to be

$$W^+ \rightarrow \mu^+ + \nu_\mu.$$

If it exists the expected 'signature' of the W in a spark chamber would be a μ^-, μ^+ pair. In the bubble chamber it is impossible to distinguish sufficiently between muons and pions to look for such muon pairs, but other modes of decay involving electrons or pions would also be expected. Among the 10 000 events in the spark chambers there were no muon pairs, nor were any other modes of decay found in the bubble chamber. It was concluded that if the W exists at all its mass must exceed 2 GeV, which is much higher than was first imagined. In many ways the absence of the W both in high-energy neutrino experiments and in other subsequent experiments has made the studies of neutrinos even more interesting and more essential for a fundamental understanding of the weak processes.

Obviously, neutrino physics will become a major experimental field, since new equipment being constructed will greatly increase the intensity of neutrino beams and the sensitivity of the detectors. New power supplies for the magnet systems of both the CERN and the Brookhaven accelerators are being installed to increase their cycling rates; the supply for the CERN machine will operate in autumn 1968. Both accelerators will have new

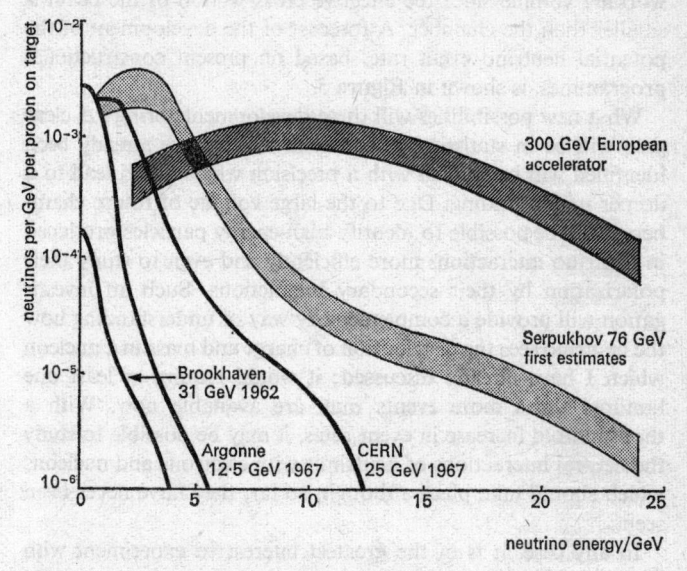

Figure 6 Neutrino fluxes produced in various machines ranging from the first high-energy experiment at Brookhaven to the 300 GeV European accelerator which may be built in the next decade are compared immediately right. If the 300 GeV were built it would create facilities for neutrino experiments very similar to those which will be provided by the accelerator to be built at Weston in the United States

injectors; that for Brookhaven will be working in 1971 and will increase the flux of the accelerated beam by about ten times. A factor of twenty in extracted proton current will be gained at each accelerator within five years and neutrino fluxes will increase accordingly. The continuous searches for more efficient neutrino-parent focusing devices will also yield a contribution.

A neutrino experiment is first priority in the experimental

programme of each of the very large bubble chambers now being built. Gargamelle, the 10 m³ chamber being constructed for CERN by the French Atomic Energy Commission, will increase the detection efficiency by a factor of seven over the chamber which has been used until now. For such large chambers the neutrino event rate is no longer directly proportional to the working volume since the effective cross-section of the beam is smaller than the chamber. A forecast of the development of the potential neutrino event rate, based on present constructional programmes, is shown in Figure 5.

What new possibilities will these developments bring? A clear gain will be in statistics; phenomena which have already been identified will be studied with a precision which could lead to a deeper understanding. Due to the large volume of future chambers it will be possible to identify high-energy particles produced in neutrino interactions more efficiently and even to study their polarization by their secondary interactions. Such an investigation will provide a complementary way of understanding how the neutrino sees the distribution of charge and mass in a nucleon which I have already discussed; it would require at least one hundred times more events than are available now. With a thousandfold increase in event rates, it may be possible to study the neutral interactions of neutrinos with electrons and nucleons which should take place although, so far, they have never been seen.

In any case, it is of the greatest interest to experiment with higher-energy neutrinos since the new phenomena they might reveal could provide essential clues to the nature of neutrino interactions. The next major advance in neutrino energy will be in the Soviet laboratory at Serpukhov where the world's newest and most powerful proton accelerator has functioned at 76 GeV since September 1967. Further developments will raise the energy to 80 GeV, perhaps even to 90 GeV. There lies the best possibility of continuing the search for the speculated W; if its mass is less than 3 GeV and its properties as predicted, it could be found at Serpukhov. The study of neutrinos has been a long standing and major interest of Soviet physicists; the invention and development of the neutrino facilities which will make the best use of this new accelerator is a fascinating challenge to experimentalists. A first estimate of the neutrino spectrum at Serpukhov is shown in Figure 6, together with corresponding spectra; from the first high-energy experiment at Brookhaven to those of Argonne, CERN and the 300 GeV European accelerator which may be built in the next decade.

It is likely that the experimental installations at Serpukhov for neutrino and other high-energy experiments will develop as a transition between the older accelerators and the 'super'-high-energy accelerators of the future. The American 200 to 500 GeV proton accelerator is already being built at Weston near Argonne. Such an accelerator will yield much higher neutrino energies and almost undreamed-of intensities. It has been quoted that if the Weston neutrino beam were aimed at the new large hydrogen bubble chamber at Argonne 25 miles [40 km] away, the chamber would record neutrino events. Clearly, the Weston laboratory will also have its own neutrino facilities and detectors. We shall probably have to wait for such facilities before it is possible to study in detail the interaction of a ν_μ with an electron which produces a μ^- and a ν_e, although it may be possible to detect it at Serpukhov.

In recent years, extensive arrays of detectors have been installed in gold mines in India and South Africa to measure the intensity of cosmic-ray neutrinos. At the depth of the experiment, only cosmic-ray muons from the vertical direction can reach the detectors; horizontal muons must be attributed to high-energy neutrino interactions in the neighbouring rocks. Indeed, the horizontal muon event rates observed are consistent with the neutrino flux, estimated from the kaon, pion and muon cosmic rays in the upper atmosphere, which are the predominant sources of these neutrinos. Until now the data which have been obtained are too few to arrive at any conclusion other than that high-energy neutrino interactions are being observed.

Cosmic rays will always contain some particles with higher energies than those produced in man-made laboratories. It could even be that the cosmic-ray installations might give an answer about the existence of the W-particle before laboratory experiments. Thus it is not only accelerator neutrino experimentalists who are planning for the future; their cosmic-ray colleagues are looking for better and more powerful detectors in their relentless search for higher event rates. One plan foresees the detection of Cerenkov light in the depths of the sea as an indication of the secondary particles from high-energy neutrino experiments. The idea is ambitious, but only by an advance of this kind, or by the proposals for the extension of the experiments in the mines or by the construction of other underground laboratories, can one hope to look for extraterrestrial sources of neutrino.

Our understanding of the universe has increased with the understanding of the forces which govern it. Because of our environment, the earliest force to be acknowledged was that of gravitation. We have learned to attribute the motion of the planets, even the formation of the stars themselves, to the role of the gravitational force on visible matter. The importance of the study of this force is unquestionable.

In the world of particle physics, the weak force plays an incomparably more important role than gravitation. If in the neutrino sea – that until now unobserved flux of cosmic neutrinos – there could be more energy than in all the visible matter of the universe, should we not devote our studies also to the all-pervading particles of that sea? As with the study of the gravitational force, it is already clear that their elusiveness in the laboratory is no measure of their fundamental importance. From the continued observation and analysis of neutrino interactions science will unravel a little more of the thread of understanding of the universe in which we find ourselves.

Part Three **The Solid State**

Solid-state physics is a very wide-ranging topic, covering as it does all aspects of matter in the solid form: the crystalline structure of matter and the explanation of the physical properties of materials; the electrical properties of conductors, superconductors and semiconductors; the magnetic properties of paramagnetics, diamagnetics and ferromagnetics. It is a subject which illustrates particularly well the interplay between many different branches of science, for amongst those concerned with problems in the solid state are theoretical and experimental physicists, mathematicians, electronic and mechanical engineers, metallurgists and crystallographers. Also it is a subject which has many, obvious practical applications. I have not included papers on these applications as they are more relevant to books on electronics or metallurgy or materials science. I have however included a paper on superconductors, another aspect of the solid state which may hold considerable promise for future development, particularly if materials can be developed which are superconducting at higher temperatures than metals.

Ziman's paper was written for a special number of *Physics Today* to mark twenty years of publication. Accordingly it reviews twenty years of progress in the subject, in a very general way, and serves to show what problems the solid-state physicist has been concerned with and what developments are likely in the future. The paper does contain a number of unexplained terminologies, most of which will at this stage have to remain unexplained, but this need not deter the student. Regard this paper as giving in broad outlines the scope of solid-state physics.

The existence of the superconducting state of metals was discovered over fifty years ago but a satisfactory explanation of this phenomenon was not put forward until much more recently. It has also been discovered in the last few years that there are in fact two types of superconductor and this is discussed in the paper by Rose-Innes which was first published in 1965.

Cottrell's paper deals with several aspects of metals, and in particular their electrical and mechanical properties. He also touches on the properties of intrinsic and impurity semiconductors.

9 J. M. Ziman

Solid State

John M. Ziman, 'Solid State', *Physics Today*, vol. 21, 1968, no. 5, pp. 53–8.

The last two decades may well have been the era of solid-state physics. Certainly it has growed and growed; absolutely, of course, like all science and all physics, but also relatively to other fields. I guess it now occupies almost 40 per cent of the family bed, instead of about 15 per cent at the end of the Second World War. With its leading American and British exponents at the head of the National Academy of Sciences and in the Cavendish Chair, even the disciples of Rutherford are being forced to acknowledge its existence!

Excuse the impudence of trying to sum up such a vast amount of human activity. Intellectual history is notoriously difficult and always false. 'No names: no pack drill' is a safe principle – and not inappropriate to a discipline that is not yet entirely dominated by the 'star' system. Solid-state physics is too diversified, in subject matter, technique and scientific motivation, to be ruled by too few big names or to be corrupted by the lure of too many Nobel prizes. Let me write, instead, about trends and fashions, movements and achievements even though these only exist in the minds of tens of thousands of research workers and in the words and symbols of a hundred thousand papers.

A golden heritage

Surprisingly our era was not a period of rampant ideological revolution. Most of the basic concepts of the modern theory were already invented by 1945. The foundations of lattice dynamics (that is, *phonons*),† of *electronic band structure*, of electron

†[We have seen how the application of quantum theory to electromagnetic waves results in the concept of photons; a similar application to elastic waves in solids and liquids leads to the concept of phonons. Ed.]

dynamics in crystals (for example, *holes* in filled valence bands) and of *spin waves* had been well laid in the 1930s as an immediate consequence of the discovery of quantum mechanics. It was already established practice to use *group theory* wherever possible; *many-body* effects associated with the electron–electron Coulomb interaction were recognized; the *Ising model* for order–disorder phenomena was familiar; the analysis of imperfect crystallinity in terms of relatively well defined and stable entities called *dislocations* was already well understood by those who could understand that sort of thing.

Has our generation invented new concepts to match the power of these? I can think of only two really big and revolutionary ideas that have been both invented and come to fruition in our time. The first would be the *quasiparticle* concept. The 'true' particles of any solid – atomic nuclei and electrons – are always interacting so forcibly that one would think that they must always merge their individuality with the crowd. Sometimes this process is so, as in lattice waves and plasma oscillations; but the solid often behaves as if it were merely an assembly of nearly independent entities with dynamical and electrical properties akin to those of ordinary particles. Something like this was divined by the pioneers, who treated metals very successfully as if full of free electrons. Many-body theory, leaning heavily on the methods of quantum field theory, has shown how to derive, justify and make quantitative this fruitful fudge.

The notion of a *pseudopotential* is not so deep but has also proved extraordinarily useful. The problem was to decouple the electrons from the fields of the ions of the crystal lattice; out of the strong must come weakness. The discovery that the algebra used in calculating electronic band structure could be rearranged to give results that did not depend very much on the detailed arrangement of the ions furnished us with a whole new kit of tools for calculating the electrical and dynamical properties of metals and semiconductors without excessive labor. This concept is actually not standing up very well to rigorous mathematical analysis, but that is not a sufficient reason for abandoning it when it works well.

Of course there has been continual technical development of the older ideas. Mathematical fictions hidden in obscure theoretical papers have become the physical realities of everyday conversation in experimental laboratories; we juggle with *phonons* and *carriers*, *magnons* and *excitons*, *dislocations*, *jogs* and *stacking faults*, and devise techniques to isolate, purify and visibly observe them. Perhaps we have already manufactured other theoretical concepts that will similarly become part of our physical 'intuition'. It is difficult to know whether simple ideas such as *flux quantization* in superfluids or intellectual subtleties such as the notion of *broken symmetry* will come out of their homesteads into wider circulation. Nor can we be sure that the highbrow apparatus of *Green's functions* and diagrammatic series, which has provided analytical justification for more naive points of view, is destined to replace our old friend the Schrödinger equation as the natural language of the subject. Such trends are often self generating. To quote Marcel Proust: 'It was Beethoven's Quartets themselves ... that devoted half a century to forming, fashioning and enlarging a public for Beethoven's Quartets, marking in this way, like every great work of art, an advance if not in artistic merit at least in intellectual society, largely composed today of what was not to be found when the work first appeared, that is to say of persons capable of enjoying it.'

Our era has certainly demonstrated that there is plenty of scope within solid-state theory for those who enjoy constructing elaborate formalisms; some who feast heavily at this table do not always repay their hosts by genuine contributions to practical knowledge. My impression is that, until recently, solid-state theoreticians were such rare birds that they were forced to converse with their experimental colleagues. By gathering in larger flocks, they begin to address their papers only to one another, and thus become as alienated from the matter of their science as schools of pure mathematics or Babylonian cuneiform.

On the other hand, mere arithmetic, however rapidly performed, has not solved many problems. The genius of the subject is judicious approximation. There are still surprisingly few situations in which the details of geometry, forces, fields and wave functions are sufficiently accurately known for a lengthy computation to pay off with good numbers. With brute force goes, as the proverb has it, ignorance. . . .

Not Greece, but Rome

This epoch was, if not intellectually revolutionary, certainly a period of enormous technical progress. A number of entirely new experimental and observational methods have grown to maturity since the war and have vastly enlarged our knowledge of the solid state.

For example, production of bulk liquid helium has transformed *low-temperature physics* from a specialized technique, practiced in only a few laboratories and concerned mainly with the exploration of the 'properties of matter' under extreme conditions, into an ordinary piece of laboratory apparatus used to study phenomena under the simplified circumstances where thermal vibrations are stilled.

Similar freedom from extraneous interference is achieved with very *pure* and *perfect crystals*. New techniques for crystal growth, such as *zone refining* and 'pulling' from the melt, were needed originally in the fabrication of germanium and silicon semiconducting devices, but they have profoundly altered the solid-state physicist's attitude to the objects of his regard; for they allow the observation of phenomena that would otherwise be completely obscured by the scattering of electrons or phonons from static impurities and imperfections.

As in all branches of science, *electron microscopy* has revolutionized the study of crystal imperfections. Until the mid 1950s, dislocations were almost unobservable hypothetical entities invented by armchair theorists to explain the complex phenomena of plasticity; seeing them being produced, running about, jogging each other, piling up, getting pinned, etc., has transformed this empirical branch of physical metallurgy into a 'first-principles' discipline.

The development of *microwave* radar during the war also had immense consequences for solid-state physics. A strong magnetic field in a resonant cavity is an ideal milieu for tickling up the spins and orbital motions of electrons and nuclei in solids, with measurable consequences of such complexity as to demand whole tomes of explanation. At first this technique was applied mainly to relatively isolated nuclei or atoms, in crystals diluted with much water of crystallization, but we have now learnt to resonate with the elementary quasiparticle excitations of the solid as a whole, such as electrons, 'holes' and 'magnons'.

Finally I would mention *inelastic neutron diffraction* as a technique that is slowly rewriting our knowledge of the dynamical motion of atoms in solids. Given a high flux of thermal neutrons, careful instrumentation and plenty of time, one can unscramble the lattice spectrum mode by mode and also look at complicated magnetic structures and spin waves. The dream of fifty years ago, that one could discover the forces between atoms by analysing lattice vibrations, is now realizable in practice.

Improvements in conventional laboratory techniques have also opened up new fields: consider, for example, the precision and sensitivity now available in infrared spectroscopy, calorimetry or X-ray diffraction. They also make orthodox solid-state physics vastly more expensive. Except with neutron diffraction, we are not yet Big Science on the Gevavolt–Megaton–Gevabuck scale, but it requires a touch of inspiration to get publishable results without apparatus in the $10 000 range. Unfortunately two modern phenomena that can be observed cheaply – positron annihilation and the Mössbauer effect – have turned out to be difficult to use quantitatively in the study of the solid state itself. Yet the field of phenomena to be explored and charted is endless: freeze a beakerful of water; can you understand everything you see?

Everest conquered

With our new experimental and theoretical techniques, what have we learnt? The most important discovery of the epoch was undoubtedly the explanation of *superconductivity*. For half a century this mysterious state has been a subject of intensive research. By the end of the 1930s many of the macroscopic properties, such as the expulsion of magnetic flux, the variation of critical temperature with magnetic field, the nature of the intermediate state, etc., were quite well known. After the war, improvements of cryogenic technique and microwave spectroscopy elucidated some further subtle properties but did not provide a basic theory. An important step was the recognition of the electron–phonon interaction as the fundamental mechanism, but the last 1000 feet could only be climbed with oxygen: field theoretical methods facilitated construction of new quantum states of the many-body system of interacting electrons and phonons, and the demonstration that these corresponded to a stable condensed state (the 'superfluid') with quasiparticle excitations of finite energy above it.

Following the success of this brilliant theory, superconductivity has been, for the past decade, the favorite son of our subject. Many new phenomena such as flux quantization, coherent tunnelling and type-2 superconductivity have been discovered, and the periodic table has been explored thoroughly for new superconductors. The theory has been so developed that I sometimes think we know more about the superconducting state of many materials than we do about the so-called 'normal' state. By a reasonable convention, the study of *superfluidity in liquid helium* at very low temperatures is considered part of solid-state physics, and this too has now been explained qualitatively, although the quantitative calculation of observable parameters in these systems is by no means easy.

An entirely unexpected discovery was that of superconducting materials that can retain their vanishing electrical resistivity in magnetic fields of ten teslas or more. Use of these materials in powerful magnets for bubble chambers, magneto-hydrodynamic rigs, thermonuclear reactors and other expensive toys is only at its beginnings. All who have had faith in serendipity and in the potential applicability of all science, however pure, have been heartened by this success, and the hunt is on for a material that remains superconducting at room temperature (for instance). Good hunting to them, say I, provided they do not call 'Tally ho!' prematurely; for wild geese have been augured in that quarter.

Making better mousetraps

Superconductivity was mysterious and only observable under extreme conditions; *semiconductivity* was well understood in principle, and quite common, so it was not nearly so closely studied! Until the famous researches that earned a Nobel prize and elucidated the role of minority carriers, the properties of p–n junctions, transistor action and other interesting effects, semiconductors were not given the attention they deserved. That heroic episode occurred at the beginning of our epoch and has been one of the main causes for the enormous expansion, in money and men, of our discipline.

There is no need to emphasize the astonishing virtuosity of the practitioners of the new art of solid-state electronics; they can shrink a camel-sized computer until it can pass through the eye of, well, a darning needle; with a tiny crystal of semiconductor, ferrite, ferroelectric or superconductor they can measure, create and control heat, light, sound and electricity; their devices are perfectly robust, infinitely flexible in use and capable of endless refinement. This is the era of solid-state physics; we do not know when it will end.

Moreover the advancement of the practical art has gone along with serious study of the fundamental physics of the phenomena involved. To construct that ideal medium in which carriers, phonons, magnons, photons and other entities interact in desirable ways, we need to know a great deal about the electronic band structure, lattice spectrum, mobility, dislocation density, impurity levels, trapping centers, etc., of the material. There is no doubt at all that the sustained effort to acquire this knowledge, required for practical use, has uncovered many new physical phenomena, clarified many hazy ideas, motivated the development of many new experimental techniques, and set a standard of scientific precision that was previously lacking in this field. Even if many of the phenomena exploited in semiconductor devices are somewhat contrived and do not tell us very much about the microtheory of the solid state, the fact that expensive factory production can depend on correct measurements of basic parameters gives us some assurance that these will be observed with care. One must surely rejoice to see one's academic pursuits so closely linked to the real world of things to make and things to do.

Making maps

Low-temperature microwave spectroscopy, perfect crystals and pseudopotentials were the recipe for another major development – the exploration of the *Fermi surfaces* of metals. As I have remarked, the idea of representing the electronic properties of a metal by the shape of the constant-energy surface enclosing the occupied states in momentum space dates back to about 1930, but attempts to calculate these surfaces from first principles were dismally unsuccessful. The trouble was that these calculations could not be checked backwards for errors and omissions because the observable properties deduced from them were only simple scalar or tensor quantities like electrical conductivity or Hall constant. I can think of only one prewar set of observations – the magnetoresistance of some metal single crystals at helium temperatures – that contains enough 'bits' of information to tell us directly about the Fermi surface.

But in the late 1950s a number of different phenomena were suddenly shown to depend very sensitively on various geometrical parameters of the Fermi surface, and hence could be transformed into new techniques of observation. Thus the *anomalous skin effect* 'measures' an average curvature, the *de Haas–van Alphen effect* an area of cross-section, the *magnetoacoustic effect* a caliper diameter, etc., of this purely hypothetical mathematical construction. It became possible, by a series of careful experiments, to build up the whole diagram on which all other electrical properties of the metal should depend.

The very first complete Fermi surface was, in fact, so surprisingly different from our expectations that some of the best theorists had to think very hard about what it signified, and had to recall some of those fundamental properties of electron states in crystals that everyone was supposed to know but had almost forgotten. The concept of a multiply connected Fermi surface was not new, but our imagination had not grasped its potentialities.

The data came pouring in and would have saturated the input channels if the pseudopotential concept had not turned up, and it was discovered that most metals had nearly free-electron Fermi surfaces after all. Nowadays the mapping of a Fermi surface is a relatively routine operation, and suitable candidate metals are getting rarer; this is a pity since various techniques by which the whole research could be conducted far more efficiently are continually invented. However, almost all these techniques only measure at the Fermi level in pure, perfect crystals, at low temperatures. There is enormous scope for further in-depth exploration of the electronic-band structure in alloys, amorphous solids, etc., with optical, X-ray and photoemission methods.

Meanwhile perhaps we have learnt to be cautious about going off on a spree of computation without sufficient moorings of experimental information. In this branch of solid-state physics we may believe with justice that we understand all the phenomena 'in principle', and yet a large technical step must be taken before one can confidently proceed to quantitative predictions.

Bush whacking

The study of *crystal imperfections* is still in a much more primitive stage of development. The dislocation concept explained the weakness of real crystals, and at the beginning of our two decades it was also shown to be the key to the mechanism of crystal growth. But without the means to observe dislocations in action, it was possible only to guess at the events within the crystal when the material was plastically deformed, hung, drawn and quartered, broken on the rack, work hardened, or allowed to creep away. How were dislocations produced and multiplied; how fast could they move; how do they interact with one another and with impurities, vacancies, interstitial atoms, grain boundaries and other types of imperfection? Some excellent hypotheses were put forward – but also some overconfident interpretations based on purely conjectural configurations and interactions.

The direct observation of dislocations by electron microscopy and other means has brought this speculative phase to an end. On the other hand, it has demonstrated the enormous complexity of the phenomena as they really are. Trying to explain a stress–strain curve by a back-of-envelope calculation may be too naïve; merely to *describe* all the creepy-crawly creatures to be seen in an electron micrograph of a thin film of stainless steel is beyond our analytical powers. In this field, as in the cognate study of *radiation damage*, we may expect only step-by-step progress, without spectacular single discoveries (although the focusing collision phenomenon is very pretty) but patiently building up a body of reliable knowledge based on sound evidence.

Fortunately there are enormous financial interests behind such studies; goodness knows how many million dollars it is worth to know what really happens to graphite in a power reactor – and metals can get fatigued at alarming rates in expensive aircraft. We may even be on the verge of another technological revolution. The next two decades may well be the epoch of 'materials science', where the deliberate design of fibrous mixes, alloys and ceramics may transform mechancial and civil engineering on the same scale as solid-state physics has transformed electrical and electronic engineering.

Unfinished business

I have discussed a few of the major achievements of solid-state physics: I could have mentioned many other topics such as *magnetism*, the properties of *ionic crystals*, the study of *surfaces* and *thin films*, the prediction and detection of *plasma modes*, the theory of *order–disorder phenomena* and so on that have developed almost beyond recognition in the past twenty years. But complacency is our vice; so let us notice that some problems have remained manifestly obdurate.

For example, the basic theory of *ferromagnetism* in metals has not significantly advanced since the war. Yes, I know that a lot of hard work has been done, many new hypotheses hypothecated, and all sorts of formalisms formulated, but we still seek a procedure for describing the d-electrons in iron, both as free, to share in the band structure, and as localized, with strongly interacting spins.

Again the attempt to calculate the *cohesive energy* of metals has scarcely been carried beyond the first column of the periodic table. It is not just a question of using the right gimmick for the exchange and correlation energy of a free-electron gas; the chemistry of ions themselves plays a role here that we scarcely comprehend. As for thermodynamic properties of *alloy phases*, these are altogether beyond the reach of current theory. Man has known for several millenia how to mix copper and tin to make bronze – but why, why, does this work so well?

Melting, now, and the ordinary *liquid state* – these are not understood in any quantitative manner. We do not have an adequate language to describe the 'structure' of amorphous materials such as glasses and ceramics; we talk glibly of 'local order', or else we construct a hierarchy of n-particle correlation functions and hope for the best. Even pure mathematicians might find 'statistical geometry' rather fun.

But much simpler properties of almost perfect crystals have never been calculated quantitatively from first principles: the thermal conductivity of sodium chloride, the electrical conductivity and thermoelectric power of copper, the superconducting transition temperature of lead, the melting point of mercury, the diffusion coefficient for vacancies in sodium. One may find recipes for calculation of most of these properties in books published before the war, but we still await a cook who can handle all the ingredients. Perhaps it is better to travel hopefully than to arrive!

J. M. Ziman

10 A. C. Rose-Innes

The New Superconductors

A. C. Rose-Innes, 'The new superconductors', *Contemporary Physics*, vol. 7, no. 2, 1965, pp. 135–51.

Introduction

The existence of superconductors, metals which suddenly lose all electrical resistance when cooled to within a few degrees of absolute zero, has been known for over fifty years. Until recently it was thought that the electrical and magnetic properties of all superconducting metals conformed to a well-established pattern, but within the last few years it has been discovered that there is another kind of superconducting metal whose behaviour is somewhat different from that of the 'traditional' superconductors.

These 'new' superconductors have special properties which make them useful in practical devices; in particular, they are used in the construction of solenoids which can generate extremely strong magnetic fields. Such fields, with strengths between fifty thousand and one hundred thousand oersted,† were previously produced only with great difficulty, and the coils used might consume more than a million watts of power. With the new superconductors these fields are produced with virtually no consumption of power other than the relatively small amount needed to refrigerate the solenoids. Very strong magnetic fields are useful in several ways, and in particular, magnetohydrodynamic ('MHD') electrical power generation and nuclear fusion power sources may become practical possibilities if very strong magnetic fields can be produced economically. Consequently many laboratories are at present engaged in trying to understand how the new superconductors behave.

However, before beginning to describe these 'new' super-

†[Oersted is a unit of magnetic field strength H. In a vacuum a magnetic field of strength 1 oersted has a flux density B of 1 gauss (10^{-4} T).]

conductors we should perhaps describe briefly the properties of the 'old' superconductors.

Superconductors are metals which, when cooled below a certain very low temperature, suddenly lose all trace of electrical resistance. The temperature at which resistance disappears is called the *transition temperature*, and is different for each metal. Not all metals become superconducting; about half the metallic elements and a large number of alloys have been found to become superconducting at various temperatures below about 20 K. For example, niobium–zirconium alloy becomes superconducting below about 12 K, pure niobium metal below about 9·5 K, aluminium below 1·2 K, but iron and gold retain their resistance down to the lowest temperatures at which measurements have been made.

Critical currents

The current in a metal is, of course, carried by electrons moving through the material, and it turns out that a superconductor will only carry a resistanceless current if the average momentum of the electrons is less than a certain value. If the current is increased so that the electrons have more than this momentum, superconductivity is destroyed and the metal reverts to its normal resistive state. The maximum current that a superconductor can carry and yet remain resistanceless is called its *critical current*. Every piece of superconducting metal has therefore a critical current which cannot be exceeded if the wire is to remain resistanceless. The critical current decreases as the temperature is raised and falls to zero at the metal's transition temperature.

Perfect diamagnetism

We speak of *super*conductors, not *perfect* conductors, and there is some significance in this distinction, because superconductors have an additional property that a merely resistanceless conductor would not have. A metal in the superconducting state has the peculiar feature that it does not permit any magnetic flux to exist in the body of the material. When a magnetic field is applied to a superconductor, resistanceless currents begin to flow on the surface, and these circulate in such a manner that they create within the material a magnetic field which is everywhere equal and opposite to the applied field. Consequently within the bulk of the metal the magnetic induction B, and hence the magnetic flux Φ, are both zero:

$$B = 0, \Phi = 0.$$

We say, therefore, that a superconductor is 'perfectly diamagnetic'. Figure 1 illustrates the difference in magnetic behaviour

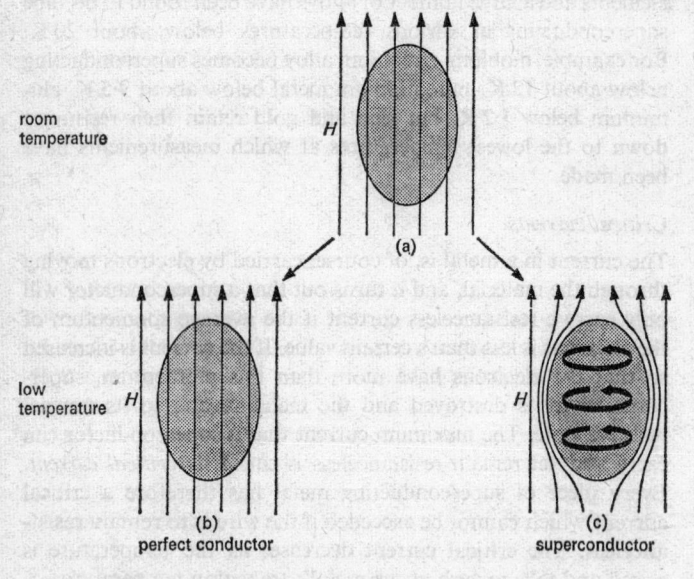

Figure 1 Difference in magnetic behaviour between a 'perfect conductor' and a superconductor

between a real superconductor and a hypothetical 'perfect' conductor. Suppose we take a piece of metal at room temperature and apply a magnetic field to it. Most metals are virtually non-magnetic and so the magnetic flux passes straight through (Figure 1a). Now cool the metal, and suppose that, below its transition temperature, the metal becomes a 'perfect' conductor. The resistivity falls to zero, but this has no particular effect on the magnetization and the flux distribution remains unaltered (Figure 1b). A *super*conductor, however, behaves differently (Figure 1c).

170 The Solid State

If we lower the temperature in the presence of a weak magnetic field, all flux is suddenly expelled from the metal when its temperature falls below the transition temperature. It can be seen that this behaviour is quite different from that of a perfect conductor.

The fact that magnetic flux is excluded has important consequences on the distribution of current flowing along a superconductor. Any electric current necessarily creates a magnetic field, and, because this field cannot exist within the superconductor, all currents must flow on the surface. In fact the currents flow in a very thin surface layer, about 5×10^{-6} cm thick. In particular, the circulating currents, which arise to cancel any magnetic field inside the body of the superconductor, also flow through this layer, and so an applied magnetic field does not cease abruptly at the boundary but falls gradually to zero within this surface layer. A magnetic field, therefore, penetrates a very small distance into the superconductor, and the thickness of the layer is called the *penetration depth*. The penetration depth is something like the *skin depth* to which high-frequency currents are confined in good conductors. We shall see later that, even though it is so small, the penetration depth plays an extremely important part in the behaviour of superconductors.

Critical magnetic field

In our consideration of the magnetic properties of superconductors we have, till now, assumed that any applied magnetic field was rather weak, because there is, in fact, a limit to the strength of magnetic field which can be applied to a superconductor. As the field strength is increased, the circulating surface currents which annul the magnetic flux within the material must also increase. Eventually, however, the critical current is reached, and when this happens superconductivity is destroyed, the metal reverts to the normal state and the magnetic flux of the applied field penetrates into the material. The magnetic field strength at which superconductivity is destroyed is called the 'critical magnetic field', H_c.

Figure 2 shows how the magnetic flux Φ through a specimen varies as we apply a magnetic field. In a normal metal the flux is simply proportional to the applied magnetic field (dotted line,

Figure 2). A superconductor, however, is perfectly diamagnetic, so, when the applied field is increased from zero, the flux is at first kept out. But when the field reaches the critical value H_c, the metal reverts to the normal state, and all the flux suddenly penetrates. At all higher fields the behaviour is just like that of any normal metal.

We noted previously that the critical current of a super-

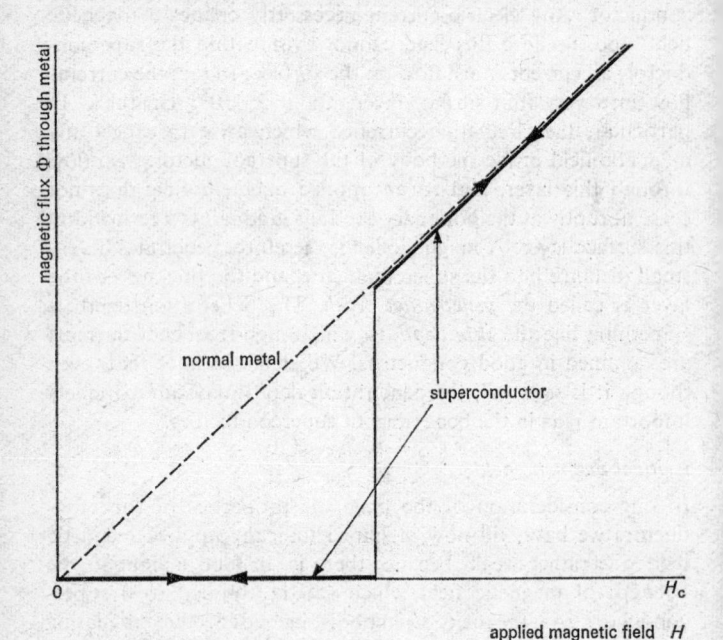

Figure 2 Penetration of magnetic flux into a superconductor

conductor falls as the temperature is raised towards the transition temperature. Consequently the critical magnetic field also decreases as the transition temperature is approached. The variation of critical magnetic field with temperature for a typical superconductor is shown in Figure 3. At a point such as P, where the temperature and magnetic field lie within the shaded region, the metal will be in the superconducting state, but (as the arrows indicate) it can be driven into the normal state by increasing either the temperature or the applied magnetic field, or both.

We see therefore that a superconductor has two possible states: the *superconducting* state which is resistanceless and diamagnetic,

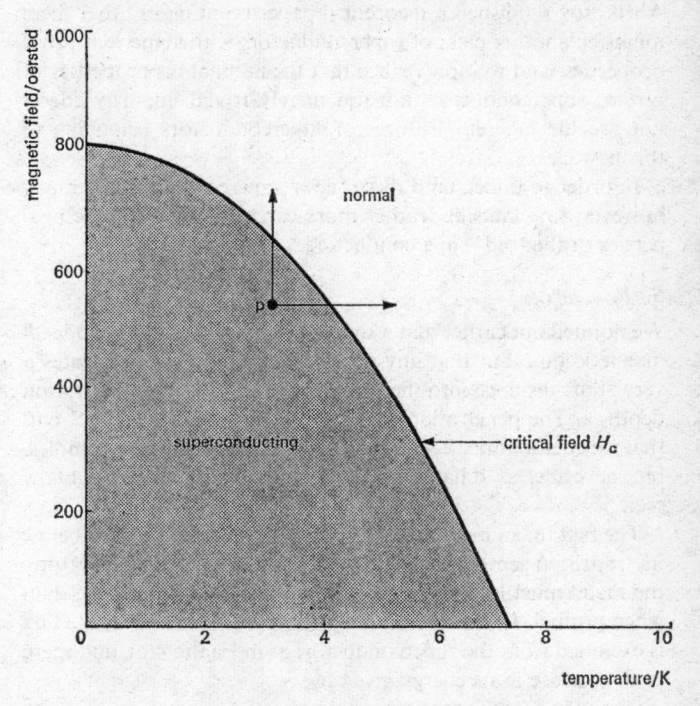

Figure 3 Variation of critical field with temperature for lead

and the *normal* state which is exactly like a normal metal. Transition to the normal state can be induced by raising either the temperature or magnetic field.

The 'new' superconductors

The features we have just described, zero resistance, perfect diamagnetism and the existence of a critical current and magnetic field, were for many years believed to be characteristic of all superconductors. It had indeed been noticed that certain superconductors, especially impure metals and alloys, did not behave quite in this way. But this anomalous behaviour was ascribed to 'impurity effects', not considered of great scientific interest, and consequently, little effort was made to understand it. However, Abrikosov published a theoretical paper pointing out that there might be another class of superconductors with somewhat novel properties, and we now realize that the anomalous properties of certain superconductors are not merely trivial impurity effects but are the inherent features of superconductors belonging to this new class.

In order to understand these 'new' superconductors we must, however, first consider rather more carefully the magnetic properties of the 'old' superconductors.

Surface energy

We pointed out earlier that a superconductor does not exclude *all* magnetic flux, but that any applied magnetic field penetrates a very short distance into the surface. We call this the penetration depth, λ. The penetration depth is so small (about 5×10^{-6} cm) that we do not notice it in measurements on normal sized samples, but, nevertheless, it has important consequences, as we shall now see.

The fact that a metal becomes superconducting if cooled below its transition temperature implies that, below this temperature, the metal must have a lower energy when superconducting than when normal. If, however, a magnetic field H is applied, the flux is excluded from the superconductor, as in Figure 4(a), and there is an increase in its energy given by

$$\Delta E_M = \frac{VH^2}{8\pi},$$

where V is the volume of the superconductor.† We might say the increase in energy results from the distortion of the magnetic flux lines. Suppose, however, that the specimen were to split into alternate thin slices of normal and superconducting phase running parallel to the applied field (Figure 4b). The magnetic

† We use the mixed c.g.s. system of units, affectionately adhered to by physicists, in spite of all international agreements to the contrary!

flux can now run down the normal slices and, if the superconducting slices are thinner than the penetration depth, the flux can also pass through them. The normal slices could be even narrower than the superconducting ones, with the result that most of the body is in the superconducting state. But, because the magnetic flux now passes through virtually undistorted, there is very little magnetic energy, and this 'mixed state' should have a lower energy than the diamagnetic state of Figure 4(a). Furthermore, because in the mixed state the magnetic energy is only a

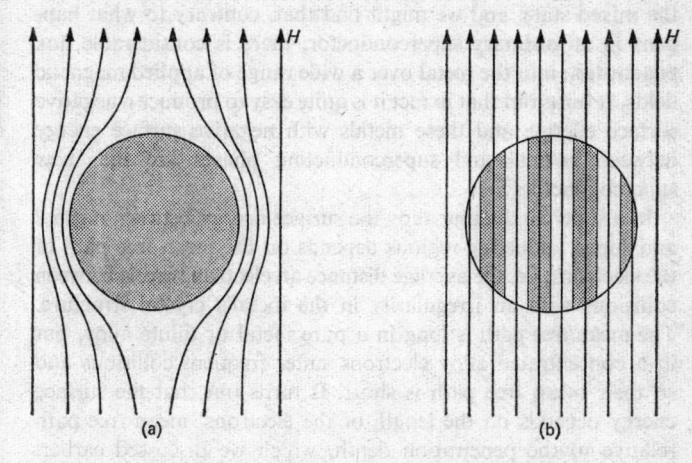

Figure 4 (a) Perfectly diamagnetic superconductor. (b) Superconductor in mixed state; shaded regions are superconducting, unshaded regions normal

fraction of $VH^2/8\pi$, this state should persist up to high magnetic fields before it becomes energetically unfavourable and the metal reverts to the normal state. We may, therefore, ask why in fact a superconductor excludes nearly all flux, even though, according to the above argument, it would seem energetically favourable for it to go into the mixed state in which flux penetrates through the metal. The answer is that, if the superconductor were to go into the mixed state, a large number of interfaces between superconducting and normal regions would have to be created. Now, a boundary between two phases always has a *surface energy*. That is to say, some energy is needed to create each unit area of interface. For most superconducting metals the surface energy between normal and superconducting regions is quite large, so considerable energy would be required to form all the interfaces of the mixed state. Hence this state is energetically unfavourable and the metal prefers to stay in the completely diamagnetic condition.

Suppose, however, we could reduce the surface energy or even make it negative (i.e. energy released when an interface is formed). It should then be favourable for the superconductor to go into the mixed state, and we might find that, contrary to what happens in an ordinary superconductor, there is considerable flux penetration into the metal over a wide range of applied magnetic fields. It turns out that in fact it is quite easy to produce a negative surface energy, and these metals with negative surface energy between normal and superconducting phases are the 'new superconductors'.

In a superconducting alloy the surface energy between normal and superconducting regions depends on the mean free path of the electrons, i.e. the average distance an electron travels between collisions with an irregularity in the metal's crystal structure. The mean free path is long in a pure metal or dilute alloy, but in a concentrated alloy electrons suffer frequent collisions and so their mean free path is short. It turns out that the surface energy depends on the length of the electrons' mean free path relative to the penetration depth, which we discussed earlier. If the mean free path is longer than the penetration depth the surface energy is positive, but as the mean free path is shortened, the surface energy is reduced, and if the mean free path is so shortened that it is *less* than the penetration depth, the surface energy becomes *negative*. (The reason for this is well understood, though rather complicated.) Hence a superconducting metal in which the electron mean free path has been reduced to less than about 10^{-6} cm should go into the mixed state which should then persist to a high magnetic field.

In Figure 4(b) we have assumed that the sample has split up into thin slices. However, in a homogeneous material there is no preferred direction in which the slices might lie and we should expect in practice to find a two-dimensional cellular pattern. Abrikosov showed that the pattern would be as shown in Figure

5. Magnetic flux, from the external magnetic field, penetrates the superconductor in a regular two-dimensional array, each flux line lying at the centre of a little resistanceless current vortex. As the applied magnetic field strength is increased the amount of flux per unit area increases (i.e. the distance between vortices decreases), so that at a sufficiently high field the flux density in the metal will be the same as that in the applied field, and the metal then returns to the normal state.

It would be interesting to test the above ideas by making a

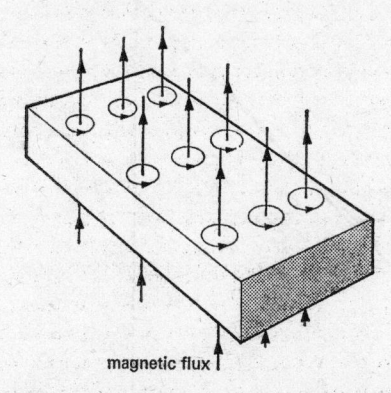

Figure 5 Abrikosov model of the mixed state, showing lattice of magnetic flux lines surrounded by supercurrent vortices

superconductor differing from an ordinary superconductor only in that its electrons have a very short mean free path. We would expect this superconductor to have a negative surface energy and go into the mixed state over a wide range of magnetic fields. It has been found that such superconductors can be produced by making suitable alloys. A particularly good example is an alloy of tantalum and niobium. Both these metals are superconductors and, because they lie next to each other in the same column of the periodic table, the electronic structures of the two metals are very similar. Furthermore, both metals crystallize in the same cubic structure with identical separation between the atoms. We may say that tantalum and niobium atoms are the same size. It is possible, therefore, to make single crystals of tantalum–niobium alloy in which the atomic sites are randomly occupied by either tantalum or niobium atoms. Because the atoms are the same size, each fits exactly into place, and we have a perfect crystal, in the sense of a completely regular three-dimensioned array of atoms. However, due to the random occupation of the sites

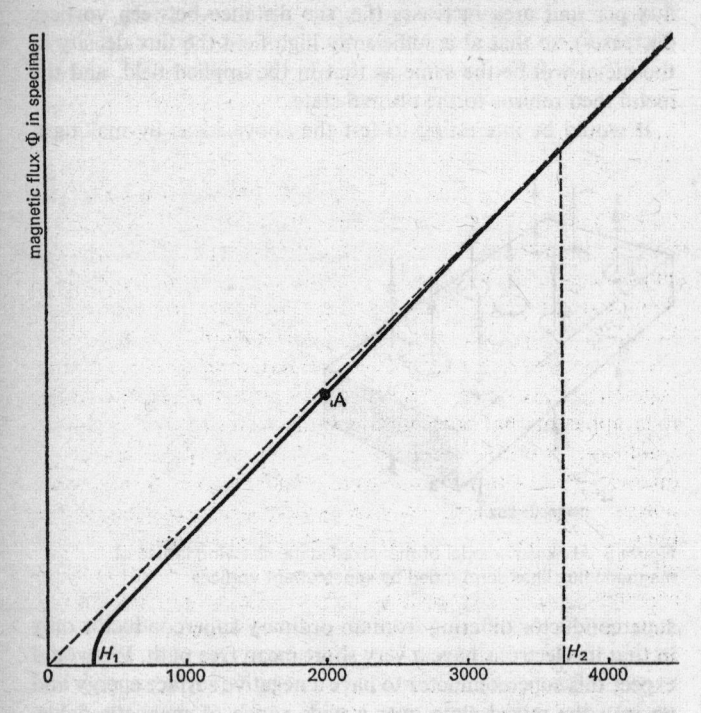

Figure 6 Flux penetration into a tantalum–niobium alloy, a superconductor with negative surface energy between normal and superconducting phases

by the two kinds of atom, the electric field pattern produced by the atoms in the crystal is extremely irregular, and electrons are scattered as they move through the crystal. Hence their mean free path is very short, which is just the condition to produce negative surface energy. Figure 6 shows the experimentally measured flux penetration into such a tantalum–niobium crystal. The flux first penetrates at a rather low applied field H_1 but the penetration is not complete until a much higher field H_2 is reached. In between H_1 and H_2 the superconductor is in the mixed state. The ability to carry resistanceless currents persists up to the upper critical field H_2. It can be seen that the form of the flux penetration curve is quite different from that of a superconductor with positive surface energy (Figure 2). At a point such as A in Figure 6 there is almost complete flux penetration even though the material is superconducting in the sense of having zero electrical resistance.

The tantalum–niobium alloy just described is the most perfect example of a superconducting alloy with negative surface energy, because the material can be grown in the form of almost ideal single crystals. But other alloys, such as lead–indium, molybdenum–rhenium and indium–bismuth, are also found to be good examples of superconductors with negative surface energy.

Because the special superconducting effects associated with negative surface energy can occur even in perfect single crystals, it is apparent that such metals form another class of superconductor. We now speak of 'type-1' and 'type-2' superconductors. Type-1 are the 'conventional' superconductors with positive surface energy between normal and superconducting phases. They have two states, normal and superconducting, and at any temperature a single critical magnetic field, H_c. Type-2 superconductors have negative surface energy and can exist in three states, superconducting, mixed and normal. The mixed state exists between two critical fields, the lower critical field H_1 at which flux first penetrates, and an upper critical field H_2 at which penetration is complete. The behaviour of the two types of superconductor is compared in Figure 7.

Any superconductor in which the mean free path of the electrons is less than the penetration depth is a type-2 superconductor, but in most alloys the foreign atoms are not the same size as the atoms of the host metal, and do not fit properly into the crystal. As a result, each impurity is surrounded by a region of distortion and the material is no longer uniform. Furthermore, most metals only have a limited solubility in each other, and, if we try to put in a lot of impurity atoms, they do not remain dispersed through the host metal, but clump together to form inclusions of a second phase. Consequently most alloys are very uneven in composition. This has important effects on their magnetic properties, because regions which are different from their surroundings act as barriers to the movement of magnetic flux lines threading the material. As a result, when we apply a magnetic field, the flux is hindered from entering the sample and at any value of applied field the flux within the sample is less than it would be if the material were pure. Moreover, when the applied magnetic field is removed, some of the flux is prevented

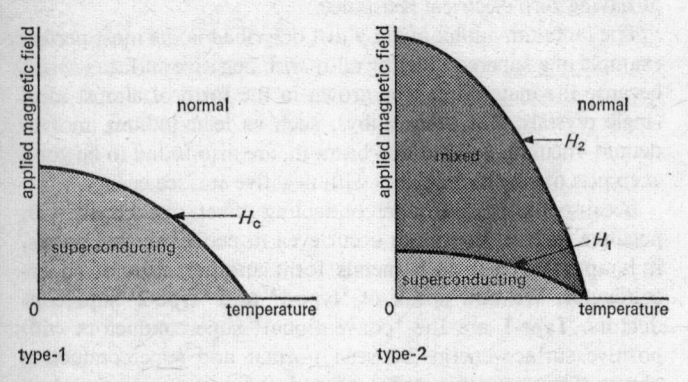

Figure 7 Phase diagrams for type-1 and type-2 superconductors

from escaping, and the superconductor is left containing *trapped flux*. In other words, the material exhibits magnetic hysteresis.

The fact that most superconducting alloys trap flux has, as we shall see, important practical consequences.

Current-carrying capacity

We mentioned that the recent intense interest in the new superconductors has arisen because they can be used to wind resistanceless solenoids producing magnetic fields up to 100 000 oersted. Since the windings have no resistance, no electrical power is consumed even when large currents are fed through them, whereas coils wound from ordinary resistive metal may consume a great deal of power. However, the ideal examples of type-2 superconductor, such as the tantalum–niobium or lead–indium alloys, though of considerable scientific interest, are not of much practical use, because their upper critical fields are only a few thousand gauss. Hence solenoids wound from these alloys would drive themselves into the normal resistive state long before they had produced very intense fields. But in some alloys and metallic compounds, especially those with a rather high transition temperature, the mixed state can persist up to fields of about 200 000 oersted, and these are the materials used to make superconducting solenoids. At present these alloys can only be produced in rather an impure form, so they are not magnetically reversible and trap large amounts of flux. We shall see, however, that, far from being a disadvantage, it is this feature which makes the materials technically useful.

It might be thought that the problem of materials for superconducting solenoids is solved if we can make superconductors in which the normal phase does not return until a very high magnetic field is applied. However this property is not by itself enough. In order to produce strong fields, we need to pass large currents through the solenoid windings. For example, a current of 30 A may be needed to produce a field of 80 000 oersted. We are therefore interested in *how much* resistanceless current a type-2 superconductor can carry when it is in the mixed state.

If, however, we measure the critical current of a wire of type-2 superconductor such as tantalum–niobium alloy, which is ideal in the sense that it has crystalline perfection, we find to our surprise that, when it has been driven into the mixed state by a magnetic field applied perpendicular to the wire, it can only carry a very small resistanceless current. (We are interested in the effect of a magnetic field *perpendicular* to the wire, because in a simple solenoid the magnetic field is at right angles to the windings.) In fact, the purer the material, the smaller the critical current. This is a rather surprising result, because commercially-available alloys such as niobium-titanium used in superconducting solenoids can carry lossless currents of up to 10^4 A cm^{-2}, producing fields of 100 000 oersted. These alloys are type-2 superconductors but, clearly, there must be some additional mechanism which enables them to pass large resistanceless currents. We shall call those superconductors which are able to carry large resistanceless currents in high magnetic fields 'high-current superconductors'. Figure 8 shows a typical dependence of critical current on applied magnetic field for a wire of high-current superconductor. At first the critical current decreases very quickly as the magnetic field is increased, but there is a plateau at about 45 A which extends up to an applied field equal to about 50 000 oersted. At magnetic fields stronger than this the ability

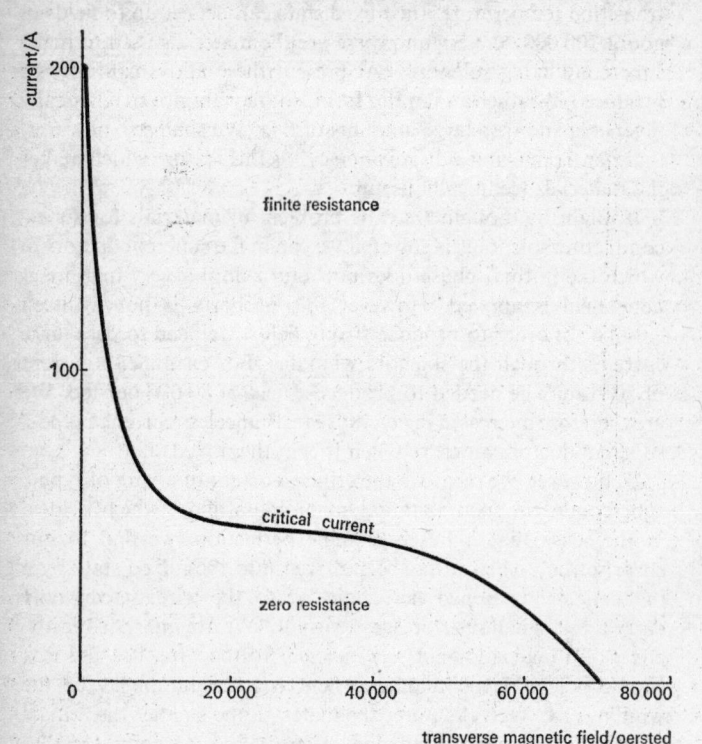

Figure 8 Variation of critical current with applied transverse magnetic field for 1 mm diameter wire of Nb–25% Zr alloy

to carry large resistanceless currents falls off. We should be able to pass 40 A through a solenoid wound from this wire and generate magnetic fields up to about 40 000 oersted.

We must now ask how it is that high-current superconductors can carry large resistanceless currents in strong magnetic fields.

The fact that they are type-2 superconductors explains why they are not driven normal until a high magnetic field is reached, but we have seen that in a very perfect example of type-2 super-

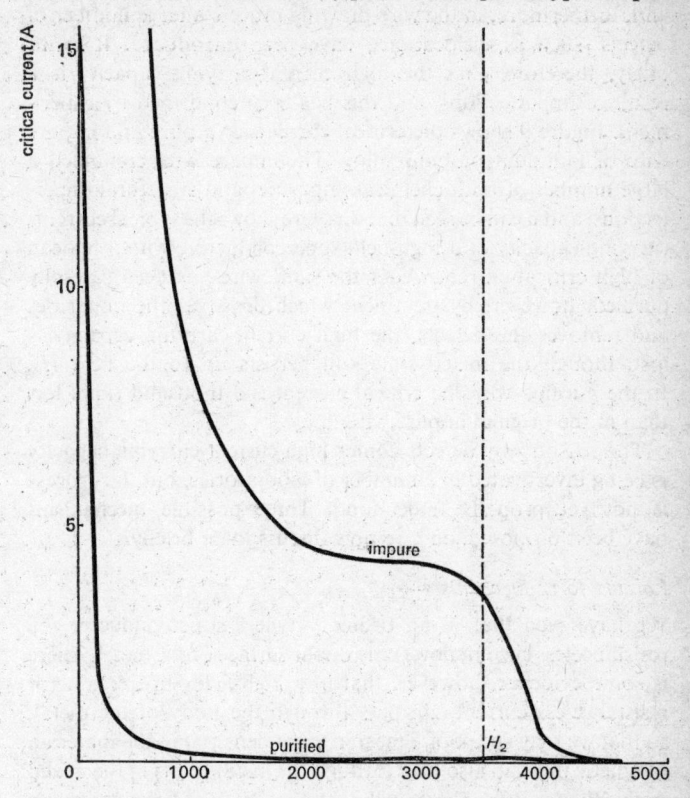

Figure 9 Critical currents of impure and pure tantalum–niobium alloy wire. (Caverley, A. and Rose-Innes, A. C., *Proceedings of the Royal Society*, Series A, vol. 255, p. 267)

conductor the mixed state is not able to carry any appreciable resistanceless current. What is the extra mechanism which confers the ability to carry large resistanceless currents?

The most obvious difference between ideal type-2 superconductors and high-current superconductors is that high-current superconducting wires are extremely inhomogeneous. They are multiphase alloys with inclusions of various phases and, furthermore, in the wire-drawing process a large number of defects, such as dislocations, have been introduced. It seems likely, therefore, that the high current-carrying capacity is a result of imperfections, and this has been confirmed by experiment. Figure 9 shows the critical currents of a pure and impure wire of tantalum–niobium alloy. The impure wire contained a large number of both chemical impurities and structural imperfections, and it can be seen that this wire shows the typical current-carrying capacity of a high-field superconductor, with a plateau of high critical current. After the same wire has been carefully purified, however, by treatment which disperses the impurities and removes the defects, the high current-carrying capacity is lost, though the mixed state still persists up to the field H_2. In the purified wire the critical current is a thousand times less than in the original impure wire.

The reason why defects confer high current-carrying capacity is being investigated in a number of laboratories, but the process is not yet properly understood. Three possible mechanisms have been proposed, and we now discuss these briefly.

Lorentz-force mechanism

We have seen that in an ordinary type-1 superconductor the resistanceless current flows only on the surface of the metal. There is some evidence, however, that in a high-field superconductor resistanceless current can pass through the body of the metal, so that we may speak of a mean current density j in the material. Magnetic flux can also pass through the metal if it is in the mixed state (Figure 10). There will therefore be an electromagnetic force (Lorentz force) between the current and the flux. The flux passes through persistent current vortices in the superconductor (Figure 5) and if each vortex encircles an amount of flux Φ, there will be a Lorentz force

$$F = j\Phi \sin \theta$$

on each vortex, where θ is the angle between the direction of the flux lines and the direction of the current. The force acts in a direction which is perpendicular to both the direction of the flux and the direction of the current. When a magnetic field is applied at right angles to the wire, as in Figure 10, the force is a maximum and tends to push the flux and hence the vortices sideways through the wire. If the vortices move and if there is opposition to their motion through the metal, work must be done to move them and, since this work must be supplied by the current, the metal appears to have resistance. Suppose, however, that impurities and defects

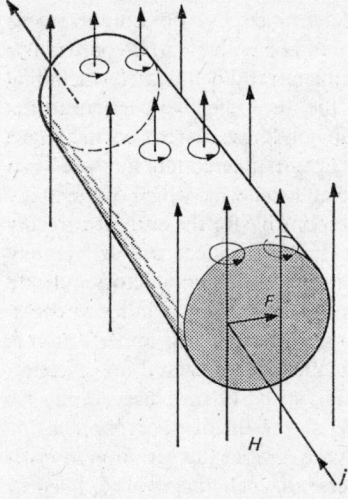

Figure 10 Type-2 superconductor carrying current through the mixed state

act as barriers to the motion of vortices. In this case the vortices will not move until the force on them is large enough to push them past these barriers. Hence, because the Lorentz force is proportional to the current density, a large current is needed to start the motion, and so the critical current will be high. In a pure homogeneous material, however, there are no imperfections to pin the vortices in place and so they begin to drift under a weak force. This would explain the small critical current in pure type-2 materials.

If the critical current is that which produces a Lorentz force strong enough to push the vortices past any barriers, we should, from the equation $F = j\Phi \sin \theta$, expect it to depend markedly on the angle between the current and the magnetic flux. In fact if we were to apply a magnetic field *parallel* to the wire, the critical current should be large because the current now exerts no force on the magnetic flux lines. Experimentally this is found to be so. The critical current of a pure type-2 superconductor in the mixed state can increase a thousandfold if the applied magnetic field is rotated from a direction lying perpendicular to the wire into a direction parallel to the wire.

If the critical current is determined by the Lorentz force mechanism just described, it should be possible to prepare alloys with a high current-carrying capacity by deliberately including those imperfections which are the most effective barriers to the movement of vortices. Unfortunately those materials which have been found to be good high-current superconductors are such complicated alloys that it is difficult to decide which of the many defects they contain is chiefly responsible for the current-carrying properties. Indeed, experiments in which defects are deliberately introduced into otherwise pure type-2 superconductors indicate that almost any imperfection is capable of pinning vortices, though it is not possible yet to say what kind of imperfection is the most efficient. It seems that the kind of crystal imperfection known as a 'dislocation' (a long string of misplaced atoms) is very effective. Figure 11 shows the result of an experiment to demonstrate this. The lowest curve A gives the variation of critical current density as a transverse magnetic field was applied to a relatively thick wire of nearly ideal type-2 superconductor, which was almost free of dislocations and other defects. As we would expect, the critical current density in the mixed state was very low: curve A, in fact, hardly has any plateau at all. The wire was then cold-drawn down to a smaller size. This reduction in area should introduce a large number of dislocations but not produce other defects. It can be seen that cold-drawing the wire so that its cross-sectional area was reduced to a third of the original area (curve B) produced a current plateau, and further reducing the area to a sixteenth (curve C) increased the critical current in the mixed state by three orders of magnitude. This experiment suggests that dislocations are effective in stabilizing the vortices against motion under the Lorentz force, but it does not show whether dislocations are more or less effective than other kinds of defect. It has been noticed, however, that in practical high-current superconductors the current-carrying properties improve when the material is drawn down into fine

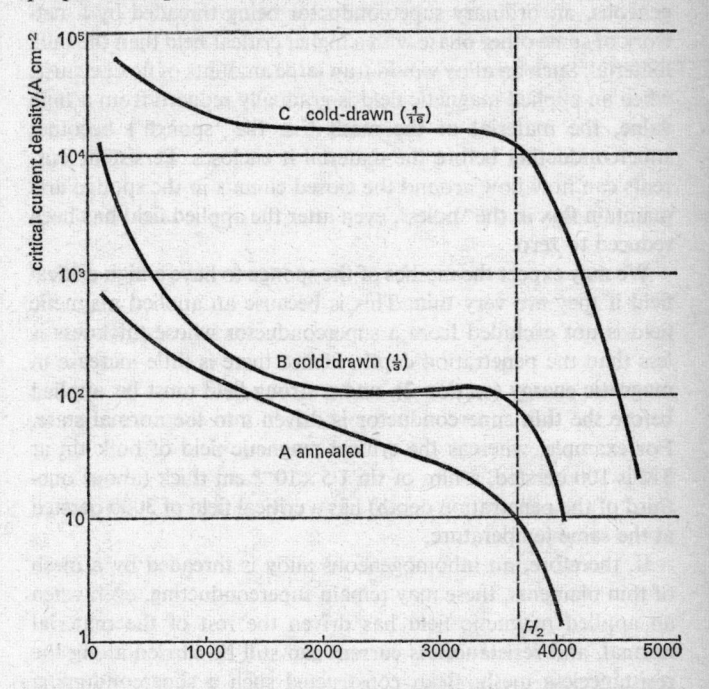

Figure 11 Increase in critical current density resulting from introduction of dislocations into a wire of pure type-2 superconductor (H_2 is the upper critical field determined by magnetic measurements)

wire, and it seems likely that the introduction of physical defects contributes to their resistanceless current-carrying capacity. In such impure, multiphase alloys, however, the effectiveness of dislocations and other crystal defects may be enhanced by impurities and phases which precipitate on them. The nature of the defects which are most effective in conferring high current-carrying capacity is still under investigation in a number of laboratories.

Superconducting 'sponge'

An alternative explanation for the properties of high-current superconductors is based on a proposal made by Mendelssohn. Mendelssohn suggested that such superconductors are inhomogeneous, an ordinary superconductor being threaded by a network of some other phase with a higher critical field than the bulk material. Such an alloy would trap large amounts of flux because, when an applied magnetic field is gradually reduced from a high value, the material of the mesh (i.e. the 'sponge') becomes superconducting before the material it encloses. Persistent currents can now flow around the closed circuits in the sponge and maintain flux in the 'holes', even after the applied field has been reduced to zero.

We may expect the meshes of the sponge to have a high critical field if they are very thin. This is because an applied magnetic field is not excluded from a superconductor whose thickness is less than the penetration depth. Hence there is little increase in magnetic energy (section 2), and a strong field must be applied before the thin superconductor is driven into the normal state. For example, whereas the critical magnetic field of bulk tin at 3 K is 100 oersted, a film of tin 1.5×10^{-6} cm thick (about one-third of the penetration depth) has a critical field of 3000 oersted at the same temperature.

If, therefore, an inhomogeneous alloy is threaded by a mesh of thin filaments, these may remain superconducting, even when an applied magnetic field has driven the rest of the material normal, and resistanceless current can still be carried along the resistanceless mesh. Bean constructed such a superconducting sponge by forcing lead into the pores of a special porous glass, so that the glass contained a mesh of lead filaments which are only about 10^{-6} cm thick. At 4·2 K this filamentary superconductor has a critical field greater than 20 000 oersted, though the critical field of bulk lead is only about 550 oersted.

A metallic alloy could contain a filamentary mesh either in the form of precipitates of a second phase or as a network of crystal faults, and the resistanceless current might be carried along this mesh rather than through the bulk material itself. This theory, however, is open to the objection that, as can be seen from Figure 11, the high-current plateau always persists up to the upper critical field H_2 of the pure superconductor, irrespective of the amount of defect present. There appears to be no reason why the upper critical field of filamentary defects should always equal that of the pure bulk material.

Surface currents

An essential feature of the mechanisms described in the previous two sections is that resistanceless current is supposed to flow through the body of type-2 superconductors when they are in the mixed state. In the Lorentz-force model the current is assumed to flow throughout the metal, whereas in the sponge model the current is conducted along filaments threading the material. Some recent experimental and theoretical results, however, suggest that resistanceless current may flow on the *surface* of a type-2 superconductor.

It has been deduced theoretically that, on any part of the surface of a type-2 superconductor which is parallel to the applied magnetic field, a skin remains superconducting, even when the strength of the field is greater than H_2. In fact, this resistanceless skin persists up to an applied field strength H_3, equal to $1·7 H_2$. An effect of the skin can be seen in Figure 11. The critical currents do not fall abruptly to zero at H_2 but small resistanceless currents can still be carried in higher fields. These resistanceless currents flow in the two narrow strips of superconducting skin on each side of the specimen where the surface is parallel to the applied transverse magnetic field.

Recent experiments have indicated that, even in applied magnetic fields below H_2 at least part of the resistanceless current is carried along any boundary of the specimen which is parallel to the field. Workers at the General Electric Laboratories have measured the critical current through a long lead–thallium alloy bar whose cross-section had the form of a triangle, and they found that, in a transverse magnetic field, the critical current was much larger when the applied field was parallel to any one of the faces. It has also been observed that the critical current of a very pure tantalum–niobium alloy depends markedly on the condition of the surface. Etching the surface of a spark-machined rod can treble the critical current. Both these experiments imply that in reasonably pure type-2 superconductors a considerable fraction at least of the current is carried on the surface. It may be that in high-field superconductors defects produce 'surfaces' inside the material (e.g. boundaries between phases, sheets of dislocations, etc.) along which the current is carried, with the result that such imperfect specimens will be able to carry more total current than ideal specimens.

Conclusion

We now know that there are two kinds of superconductor: type-1 superconductors which have only two states, superconducting and normal, and the new type-2 superconductors with three states, superconducting, mixed and normal. The mixed state is resistanceless but, unlike the superconducting state, flux from an applied magnetic field penetrates through it. In consequence, the mixed state persists up to high magnetic fields, but, if the material is pure and homogeneous, only a small resistanceless current can be carried.

The presence of imperfections and defects enables much larger currents to be carried, and such imperfect type-2 superconducting metals are used to wind solenoids capable of producing magnetic fields as strong as 100 000 oersted. We do not, however, at present understand why defects confer this high current-carrying capacity on type-2 superconductors, though several mechanisms, Lorentz-force, sponge and surface conduction, have been suggested. It does not, of course, follow that the current-carrying capacity of practical high-field superconducting alloys is a result of any one single mechanism. It may well be that in complicated multiphase alloys more than one of the above mechanisms is contributing to high resistanceless current.

With regard to high-field superconducting solenoids, we are in the not uncommon technical situation of using a device without properly understanding how it works. Though such a situation is better than its converse, it makes it difficult to produce improvements. For this reason there is considerable effort in many laboratories to try and understand the behaviour of these new superconductors, so that we may learn how to produce better alloys and material which is more predictable in its properties.

11 A. H. Cottrell

The Metallic State

A. H. Cottrell, 'The metallic state', *Contemporary Physics*, vol. 1, 1960, no. 6, pp. 417–32.

1 Introduction

The properties of metals, such as mechanical strength, ductility, electrical conductivity and ability to combine in various proportions to form alloys, are so remarkable and so valuable to mankind that a special branch of physical science has grown up to deal with them. It is the task of the metallurgist to develop these and other properties to greatest effect in practical materials by placing the right atoms together in the right amounts and in the right positions. Great changes have taken place in metallurgical science in recent years and, as a result of our much better understanding of the metallic state, it is becoming possible to design metals and alloys systematically from a knowledge of atomic properties. The constant interplay of pure and applied science provides one of the fascinations of modern physical metallurgy.

An ordinary piece of metal usually consists of many small polyhedral crystal grains completely joined together by grain boundaries, these being transition regions two or three atoms thick across which the crystal lattice changes its orientation from one grain to the next. It is general practice to make the grains small, e.g. about 0·1 mm across, by various alloying, casting, working and other treatments, because a small grain size gives good mechanical properties, but it is not difficult to grow large single crystals when necessary. The atoms in metal crystals are usually arranged in strikingly simple patterns, such as may be produced by stacking equal spheres together in close-packed formations. Figure 1 shows the three most common crystal structures, with the atomic spheres contracted to points for clarity.

The close packing of the atoms in these structures is indicated by the high 'coordination number', the number of nearest neighbours to an atom. In a face-centred cubic metal, for instance, this is twelve. Such high coordination numbers obviously cannot be explained by ordinary notions of chemical valency. The well-known rule that an atom of valency N can bond with $8-N$ neighbours, which works so well in, for example, tetrahedrally

face-centred cubic
(Cu, Ni, Al, Pb, Fe at high temperatures)

close-packed hexagonal
(Mg, Zn, Cd, Be)

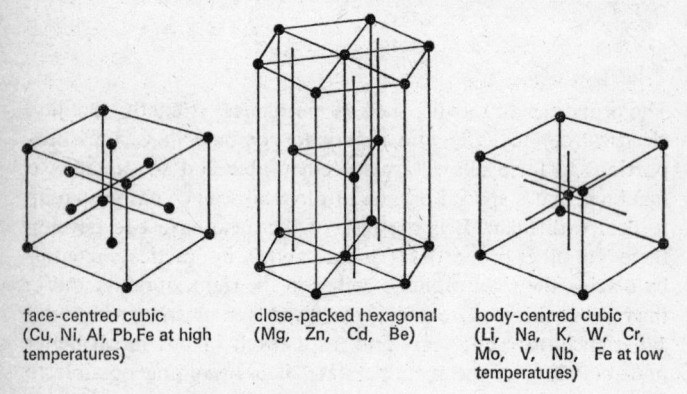

body-centred cubic
(Li, Na, K, W, Cr, Mo, V, Nb, Fe at low temperatures)

Figure 1 Crystal structures of metals

bonded diamond, clearly breaks down completely when applied to metals.

2 Free electrons in metals

Metals differ from non-metals most strikingly in their electrical conductivities, those of copper, silver and aluminium, for example, being 10^{20} times larger than those of diamond, mica and aluminium oxide. Drude and Lorentz explained this by postulating the existence in metals of 'free electrons', that had become dissociated from particular atoms in the metal and moved freely as a kind of gas or 'plasma' through the entire network of atoms. This idea has dominated the whole subject ever since. As we shall see later, however, it is not the ability to conduct electricity by the passage of free electrons that distinguishes metals from insulators, for the same process can be made to occur in an insulator. The special feature of a metal is that electrons in it remain free right down to the lowest possible temperatures. In fact, the electrical conductivity of a metal improves on cooling to low temperatures whereas that of an insulator or 'semiconductor' decreases towards zero.

The Drude–Lorentz theory has been extremely successful. Electrical and thermal conductivities, and optical properties, were explained in terms of the movements of free electrons in electrical and thermal gradients. Tolman and Stewart proved that sudden pulses of current could be set up by shaking metals sharply, so that the free electrons were thrown towards one end.

Measurements, particularly of electrical conductivity, showed that in a metal such as silver there is approximately one free electron per atom. Considered as a gas, the electronic fluid is thus extremely dense and it is not surprising to find that it behaves in a radically different manner from other gases. Sommerfeld showed that this high density leads to important quantum effects, particularly on the thermal properties of the electron gas. To describe a gas by the classical kinetic theory it must be physically possible, in principle at least, for us to separate individual particles of the gas. If n is the number of free electrons per unit volume then we must be able to locate the position of any electron with an error $(\Delta x)^3$ smaller than n^{-1}. But we know from Heisenberg's uncertainty relation that a particle cannot be located with a precision Δx unless its momentum exceeds a minimum value Δp given by

$$\Delta p \, \Delta x \simeq h, \qquad \qquad 1$$

where h is Planck's constant ($6 \cdot 627 \times 10^{-34}$ J s). Since the energy E is related to momentum p and mass m through $E = p^2/2m$, this means that the gas cannot behave according to classical kinetic theory until it has been raised to such a temperature that the thermal energy per electron exceeds the value

$$E_0 \simeq \frac{(\Delta p)^2}{2m} \simeq \frac{h^2 n^{\frac{2}{3}}}{2m}. \qquad \qquad 2$$

The exact result is $E_0 = (h^2/8m)(3n/\pi)^{\frac{2}{3}}$, which corresponds to a temperature of 60 000 K for silver. At ordinary temperatures the motion of the electrons is completely dominated by quantum effects. Even at 0 K the electrons are in motion with speeds of order 10^7 cm s^{-1} and they occupy a band of quantized kinetic energy levels up to a maximum at E_0. This state of affairs persists almost unaltered at all temperatures up to the boiling point.

We must now ask why some substances are insulators and others metals. There are two approaches to this problem. The older one takes the existence of free electrons as a starting assumption and then sets out to prove that under some circumstances they can transport a current and under others they cannot. This theory has the advantage of being suitable for detailed mathematical treatment, but, as Mott has pointed out, it predicts to be metals some substances that are known in fact to be insulators. It would seem to have chosen not quite the right criterion for deciding between metals and insulators. The other and more recent approach starts with insulators and considers the conditions under which their electrons may become free. We shall follow this second approach, believing it to be intrinsically sound even though it is not yet capable of exact mathematical treatment and provides us with only a blunt-edged criterion.

3 Ionization

Matter is normally electrically neutral. An electron of charge $-e$ is attracted to a proton of charge $+e$ at a distance r by a Coulomb force $e^2/4\pi\varepsilon_0 r^2$,† and the potential energy $V(r)$ of the electron in the field of the proton is given by

$$V(r) = \frac{-e^2}{4\pi\varepsilon_0 r}. \qquad 3$$

A neutral hydrogen atom exists when the electron is captured in this field and moves in a 'bound quantum state' around the proton. Ionization into mobile, free, charged particles is essential for electrical conductivity. This occurs most obviously in hot gases, when the temperature T becomes sufficent to ionize a large number N of the atoms or molecules, according to the formula

$$N = \text{constant} \times \exp\left[\frac{-I}{2kT}\right], \qquad 4$$

†[In the original the formulae were in unrationalized c.g.s. units and I have altered these. Ed.]

where I is the ionization energy and k is Boltzmann's constant. At low temperatures the gas is an insulator, then it behaves like a semiconductor as its conductivity increases with temperature, and finally it becomes a fully ionized plasma at sufficiently high temperatures.

The ionization energy for the removal of the first electron from a gas atom varies from 4–5 eV for alkali metals to 10–25 eV for halogens and noble gases (1 eV = 1 electronvolt = $1 \cdot 6 \times 10^{-19}$ J). The total energy E of the electron in a hydrogen atom can be written as the sum of potential and kinetic energies

$$E = -\frac{e^2}{4\pi\varepsilon_0 r} + \frac{p^2}{2m}. \qquad 5$$

The electron tends to move towards the nucleus and so reduce the potential energy; but this effect is resisted by the kinetic energy, which rises as the space in which the electron moves is decreased. Heisenberg's principle shows that if the electron moves at a radius r from the nucleus its momentum cannot be smaller than $h/2\pi r$. Substituting this for p in **5** and minimizing E with respect to r, we obtain

$$r = \frac{\varepsilon_0 h^2}{\pi m e^2} = 5 \cdot 3 \times 10^{-11} \text{ m} \qquad 6$$

and $\quad E = \dfrac{-me^4}{8\varepsilon_0^2 h^2} = -13 \cdot 6 \text{ eV} \qquad 7$

for the radius and energy of the lowest electronic state of a hydrogen atom; the ionization energy is $-E$.

As regards more complex atoms we recall that the distribution of electronic charge about a nucleus is best described by means of 'wave functions' and that the principal quantum number n of such a function determines the radial distribution of the charge and hence mainly determines the energy of electrons in the 'shell' of quantum states with that value of n. A striking feature of atomic structures is the formation of spherically-symmetrical filled shells of eight electrons. Because of the spherically symmetrical way these electrons are disposed about the nucleus each electron in such a shell is about as near to the nucleus as any other and hence enjoys the electrostatic attraction of the nucleus with the least possible shielding or 'screening' by its companions in that shell. The ionization energy is then large. Such structures occur in atoms of the inert gases. If the nuclear charge is reduced by one unit a 'hole' corresponding to one missing electron appears in the otherwise spherical shell of the neutral atom. The ionization energy to remove one of the electrons in the shell remains high, but the atom now has the ability to attract one extra electron, so becoming a negative ion, to fill the hole and complete the spherical shell. Such atoms belong to the electronegative halogen family of elements. Conversely, with an alkali metal atom, which has one more nuclear charge than that of an inert gas, there is no room in the filled spherical shell for its extra electron and this is forced out into a quantum state lying further from the nucleus. So far as this electron is concerned, the effect of the extra nuclear charge is more than offset by the screening effect of the electrons in the filled shells which lie between it and the nucleus. It is thus easily ionized. The atom is then turned into a singly charged positive ion.

The affinity of electropositive and electronegative elements for one another, and the formation of ionic compounds and solids such as rock-salt, can thus be explained in terms of the transfer of electrons from one to the other and the electrostatic attraction of the ensuing positive and negative ions. The chemical valence of an atom is equal to the number of electrons which must be added or removed to bring it to the nearest filled-shell structure. When this is achieved the atom is 'chemically saturated'.

4 Insulators, semiconductors and metals

In covalently bonded substances each pair of neighbouring atoms shares two valence electrons and each atom bonds with enough neighbours to obtain the equivalent of a filled shell from these shared electrons. The atoms in diamond, silicon and germanium have four neighbours each and their chemically saturated bond structure can be represented by a diagram of the type shown in Figure 2, in which each pair of lines represents an electron-pair bond.

The electrons in these bonds cannot transport an electrical current without disarranging the bond structure. When it is disarranged, however, for example by photoionizing one of the valence electrons with a γ-ray, a pulse of current can be produced through the crystal; this effect is exploited in the 'diamond-crystal counter' of nuclear physics. The insulating properties of diamond must therefore be ascribed, not to immobility of the valence electrons, but to the resistance of the bond structure to

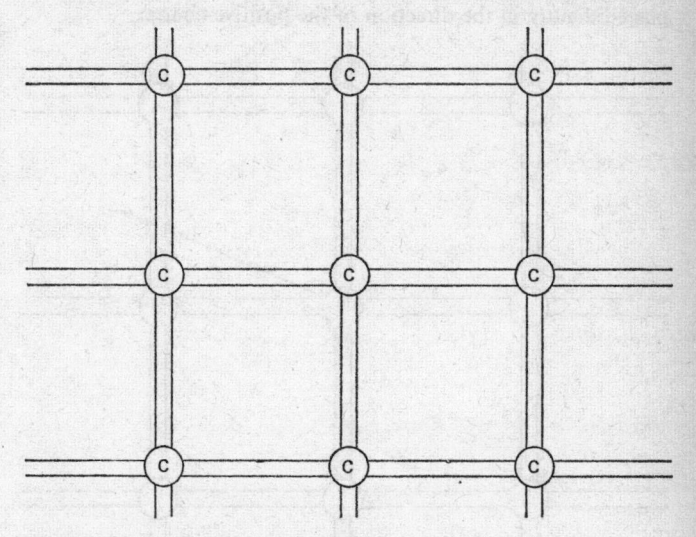

Figure 2 Covalent bonding in diamond

disarrangement. Electrons can change places with one another but this does not produce a flow of current.

As in gases, conduction can also occur through thermal ionization. Consider, for example, the behaviour of an impurity atom such as phosphorus built into the lattice of, say, silicon, as in Figure 3. Four of its valence electrons enter covalent bonds, but the fifth remains unbonded and is held electrostatically to the positive ion like the electron in a hydrogen atom. 6 and 7 can be applied to describe this electron provided we allow for the effects of the crystalline field of balanced electrical forces which forms the 'space' in which it moves. This is done in the calculations by replacing the actual mass of the electron by an 'effective mass' m^*, a parameter that takes account of the fact that an electron moving through a periodic electric field does not have the same mobility as one moving through a smooth field, and by replacing the electrostatic interaction $e^2/4\pi\varepsilon_0 r$ with $e^2/4\pi\varepsilon_0 \varepsilon_r r$. Here ε_r is the relative permittivity of the material, which allows for the fact that the covalently bonded atoms become polarized in the field of a free charge, their electrons being displaced slightly in the direction of the positive charge.

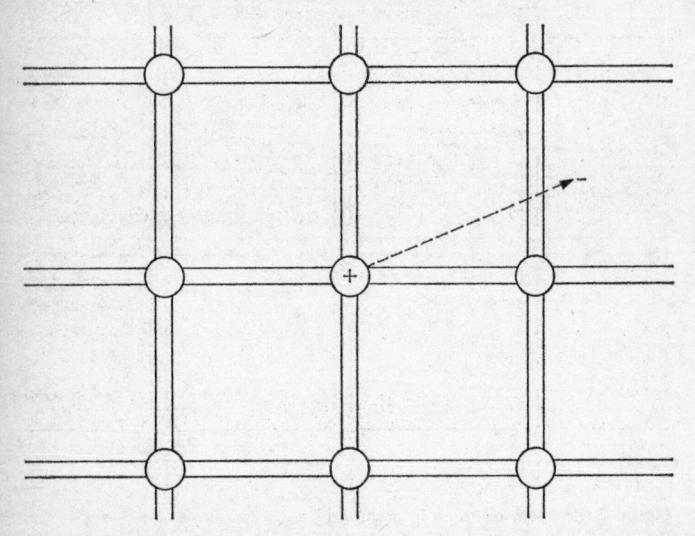

Figure 3 Representation of a phosphorus atom in silicon as an unbonded electron and a positive ion

In silicon $m^* \simeq 0.25m$ and $\varepsilon_r = 11.9$. **6** and **7** then give $r \simeq 25 \times 10^{-8}$ cm and $E \simeq -0.025$ eV. (In practice E is raised to nearly twice this value by subsidiary effects.) Due mainly to dielectric polarization the electron moves in a much larger orbit than in the hydrogen atom and is easily ionized. Ionization involves removing the electron far from its 'donor' atom and so freeing it to take part in conduction. The electron is then said to have joined the 'conduction band' of the material by the acquisition of the ionization energy, denoted in this case by E_d. Because E_d is small thermal ionization occurs appreciably even at room temperature. The material is not a metal however, because the conductivity disappears at low temperatures (cf. 4), but is an 'impurity semiconductor'.

Intrinsic semiconductivity, produced by the thermal ionization of electrons out of valence bonds, is also possible but requires a larger energy, E_g, made up of the excitation energy required to unbond a valence electron and of the electrostatic energy required to pull the electron away from the 'positive hole' which it leaves behind in the valence bond structure. Some values of E_g (in eV) are as follows: diamond, 6·5; silicon, 1·1; germanium, 0·7.

Such effects as these are usually represented by an energy-band diagram, as in Figure 4, in which energy is measured vertically

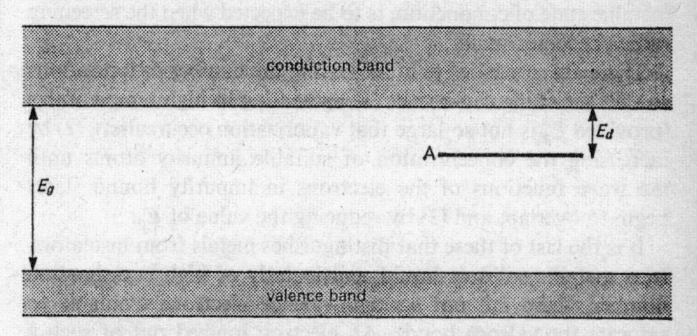

Figure 4 Bands of electronic energy levels in a covalent solid. A localized level representing the bound state on a donor impurity atom is shown at A

and distance horizontally. These bands are analogous to energy levels in free atoms; broadening into bands occurs because the closely spaced atoms interact with one another. An electron bound to a donor impurity atom is localized in a quantum state, belonging to that atom, which lies at a level E_d below the conduction band.

When large numbers of ionizations occur a new major effect sets in to accentuate further the degree of ionization – the material becomes too good a conductor to permit long-range electrical fields to exist inside it. The effect of this is rather like dielectric polarization but much more drastic; and the Coulomb interaction $e^2/4\pi\varepsilon_0 r$ is replaced, not by $e^2/4\pi\varepsilon_0 \varepsilon_r r$, but by a 'screened charge' interaction, given by

$$V(r) = -\frac{e^2}{4\pi\varepsilon_0 r} \exp(-r/r_0), \qquad 8$$

where r_0 is of the order of magnitude of the spacing between the ionized electrons. Unlike $e^2/4\pi\varepsilon_0 r$ this is a short-range interaction which virtually disappears at distances larger than r_0. When r_0 is sufficiently small (e.g. of order of the atomic spacing) the screened potential is too weak to produce a bound state for an electron at all, and the ionization energy, which is the factor that prevents electrons becoming free at low temperatures, then disappears. As Mott has emphasized, this means that a sharp increase in the number of free electrons irrespective of temperature, leading to a metallic state of conduction, is to be expected when the screening radius becomes small.

There are several ways of increasing the number of ionizations so that screening can occur: (1) by heating to high temperatures (provided E_g is not so large that vaporization occurs first), (2) by increasing the concentration of suitable impurity atoms until the wave functions of the electrons in impurity bound states begin to overlap, and (3) by reducing the value of E_g.

It is the last of these that distinguishes metals from insulators. In a metallic solid or liquid, particularly of high coordination number, there are not enough valence electrons available to saturate the valence bonds. An electron ionized out of such a bond will, wherever it goes in the metal, always be able to form a similar bond in its new position since there are vacant bonding states everywhere. The unbonding energy which contributes so strongly to E_g in a chemically saturated material such as diamond is now absent. Every atom can behave as if it were a donor impurity, offering its easily removable valence electron(s) for ionization, and the ionization energy is eliminated by the intense screening which sets in. Conditions thus favour the complete disappearance of bound states for the valence electrons. The gap between the conduction and valence bands disappears and large numbers of free electrons exist at all temperatures. This is the metallic state.

5 Effect of a periodic crystalline field

The above theory of the difference between metals and insulators is due essentially to Mott, and is closely related to a theory due to Pauling. The crystallinity of the atomic structure plays little part in the argument. Since the distinction between metals and insulators persists in the liquid state, this is reasonable.

It has often been thought that crystallinity plays an essential part, however. There is a well-established theory (the 'zone' theory) due to Bloch and Wilson which says that, because the electric field through a crystal is *periodic*, certain states of motion must be forbidden to electrons in crystals. If an external beam of electrons is allowed to impinge on a crystal, with a direction and momentum corresponding to one of these forbidden states, it is rejected by the crystal by reflection off a set of crystal planes. This is the basis of the phenomenon of electron diffraction. The forbidden states occur in certain energy bands so that energy diagrams similar to that of Figure 4 may be deduced purely from considerations of crystal symmetry and the periodic variation of potential through the lattice. A distinction between metals and insulators can then be drawn according to whether or not the allowed energy bands are partly filled by electrons.

The periodicity of the lattice field certainly does influence the behaviour of electrons in solids. Measurements of various physical properties have given evidence of the variation in the number of allowed quantum states and in the effective mass of an electron for different directions of motion through a crystal. Pippard has shown that these effects are strong even in such an 'ideal metal' as copper. Evidence for crystallographic restrictions on the motion of free electrons is also provided by the behaviour of a poorly conducting metal such as bismuth; its conductivity increases sharply on melting towards a value more typical of metals generally. It is also known that many alloys of the type described as 'electron-compounds' (e.g. $CuZn$, Au_5Zn_8, $AgCd_3$) form particularly at certain critical ratios of the numbers of free electrons to atoms (1·5, 1·61, 1·75) and that these ratios can be explained in terms of the effect of crystal structure on the energies of such electrons.

Important though these effects of the crystal structure are, one feels that they are subsidiary effects. Although insulators improve their conductivity on melting, they do not usually become metals in spite of the destruction of the long-range crystal structure on which the zone theory depends. The importance of ionization, rather than partly filled energy bands, as the criterion for distinguishing metals and insulators is shown by considering materials such as nickel oxide. Nickel is a transition metal and in forming NiO by transferring two of its electrons to oxygen to form an ionic crystal, it must leave two places vacant in its outermost electronic shell (the 3d shell). In the solid the energy levels of this shell broaden out into a band which is only partly filled. This band would therefore be expected to provide conductivity on the zone theory whereas in fact nickel oxide is an insulator.

6 Cohesion in metals

The picture we have then of a simple (non-transition) metal is one of essentially spherically symmetrical positive ions floating in a sea of mobile free electrons. The free electrons are provided by the easily ionized valence electrons of the atoms, those electrons belonging to the filled shells requiring too much excitation energy to be lifted out of their shells into the conduction band. Since there is no saturation of the bonding states for the valence electrons the $8-N$ rule no longer restricts the number or type of neighbours which an atom may have. Close-packed structures and free intermingling of different metallic atoms in alloys are thus expected. The bond that an atom makes with any single neighbour is weak but this is compensated by the stacking of the atoms in close-packed formations so that each has many neighbours and the overall cohesion is strong. Thus metals compare well with ionic and covalent solids as regards melting points, boiling points, elastic constants, and other cohesive properties.

This tendency towards close-packing leads in pure metals to the simple crystal structures of Figure 1. In certain alloys such as *Laves phases*, e.g. $MgCu_2$, KNa_2, advantage is taken of the difference in the sizes of the participating atoms to form more complex crystal structures of even closer packing in which the coordination number exceeds the maximum (12) possible for equal-sized spheres. The ability of metals to dissolve very small atoms (e.g. H, N, C) interstitially, i.e. in the spaces between the metal ions, which is of great technological importance in the case of carbon steels, is another example of this effect of the unsaturated metallic bond.

The flexibility of the metallic bond influences the surface and elastic properties of metals. If a metal is cut into two pieces along some surface the electrons previously responsible for the bonds through that surface can transfer themselves to bonds between atoms in each of the new surfaces so created. The surface energy of a metal is thus smaller than would be expected purely from the number of bonds cut in making a new surface. A similar argument holds for elastic constants. There are two basically different ways of deforming a solid elastically; by change of volume at constant shape (dilatation) and by change of shape at constant volume (shear). Resistance to the first is measured by the bulk modulus of elasticity, that to the second by the rigidity modulus. The flexibility of the metallic bond allows atoms in metals to slide over each other fairly easily and the ratio of rigidity modulus to bulk modulus is lower for metals than for most other solids. A convenient measure of this is provided by Poisson's ratio, the ratio of lateral to longitudinal strain when a bar is stretched in simple tension. Lateral contraction decreases the dilatation but increases the shear. Poisson's ratio for a liquid with no resistance to shear is 0·5; for the elastic deformation of a metal it is usually about 0·33 whereas for diamond, glass and ionic solids, it falls generally in the range 0·2 to 0·3.

To discuss the cohesion of metals quantitatively it is better to think of it in physical terms, as the coherence of positively charged ions to a negatively charged fluid, rather than chemically, as the direct bonding of atoms to one another. When free atoms condense together to form a metal their valence electrons become free to explore the whole material. The localization of such an electron to a particular atom disappears, and with it disappears the kinetic energy due to this localization (cf. **5**). The potential energy on the other hand is not much altered since the electron is never far from a positive ion. In a free atom of an alkali metal the potential energy (negative) is about twice the magnitude of the kinetic energy (positive) so that, when the atoms condense together, it is possible for an electron practically to double its binding energy. The actual increase in sodium due to this effect is 3·13 eV.

Most of the valence electrons do not do as well as this, however. There are two opposing effects. First, because these electrons all share a common space, the Pauli principle restricts them to two electrons (with opposite spins) in a quantum state of this space. The result is as described in Sommerfeld's theory (section 2). The electrons occupy all states in a band of levels up to a maximum, called the 'Fermi level', which is 3·12 eV for sodium. For an electron at the top of this band, therefore, the Fermi energy almost exactly cancels the gain in binding energy. However, the average Fermi energy of the electrons is only 0·6 of this maximum value, i.e. 1·97 eV. There is thus a net binding energy of $3·13 - 1·97 = 1·16$ eV, which is strikingly close to the measured vaporization energy of sodium (1·13 eV per atom).

The Pauli principle keeps the electrons out of one another's quantum states but does not otherwise prevent them from occupying the same point in space. Since electrons interact with one another electrostatically with a long-range potential $e^2/4\pi\varepsilon_0 r$ it would seem that their energy of interaction might make a strong contribution, additional to the Fermi energy, opposing the cohesion. In fact, early attempts to develop a free-electron theory of cohesion ran into great difficulties at this point. The effect, however, turns out to be almost negligible. The long-range part of the interaction can be shown to lead to a coherent oscillation of the whole electron gas, called a 'plasma oscillation'. This is a very interesting phenomenon, but it hardly alters the cohesive energy. When the part of the interaction represented by these plasma oscillations has been accounted for there remains only a short-range interaction very similar to that of **8**. What happens, in effect, is that the electrons correlate their positions with one another so that each carries about with it a 'positive hole' in the electron distribution in which another electron is unlikely to be found. If it were possible to measure it, the average density of electronic charge along a line through such an electron would look rather as in Figure 5. The central spike represents the electron and the area beneath it is balanced by the reduced area beneath the depressed region of the electron distribution nearby. As regards cohesion, the result for a monovalent metal is that when a given electron happens to lie on a given ion there is hardly every another free electron on that ion. There are, of course, free electrons on neighbouring ions but their electrostatic interaction with the electron in question is almost exactly cancelled by the interaction of these other ions with it. The only important interaction of the electron is that with the ion on which it happens to lie, as in a free atom.

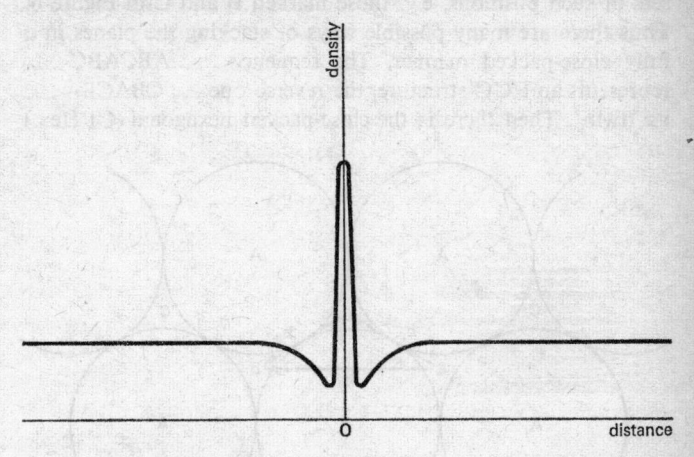

Figure 5 Expected variation in density of electronic charge along a line through a free electron

7 Close-packed structures

It is a long step from free electrons to the engineering properties of metals, but the shortest route lies through the metals copper, silver and gold. In these the free-electron gas pulls the ions together until the filled electron shells of neighbouring ions begin to overlap and exert a strong resistance to further interpenetration. This ion–ion repulsion, which is mainly responsible for the observed lattice constants of these metals, is a short-range 'central' force, i.e. it acts between nearest neighbours along the line between their centres. The free electrons, by contrast, are sensitive essentially to the *volume* allotted to each atom rather than to the distance between neighbours.

The need to keep the volume small and the ionic spacing large is best satisfied when the atoms are packed in a structure of maximum coordination number. For a plane sheet of equal spheres there is only one such structure, as represented by the circles centred at positions A in Figure 6. The face-centred cubic

(FCC) crystal structure can be built up by laying several such planes on top of one another so that the atoms of one rest in the hollows of another. However, there are always two alternative sets of such positions, e.g. those marked B and C in Figure 6. Thus there are many possible ways of stacking the planes in a fully close-packed manner. The sequence ... ABCABC ... represents an FCC structure, the reverse one ... CBACBA ... its 'twin'. Then there is the close-packed hexagonal (CPHex.)

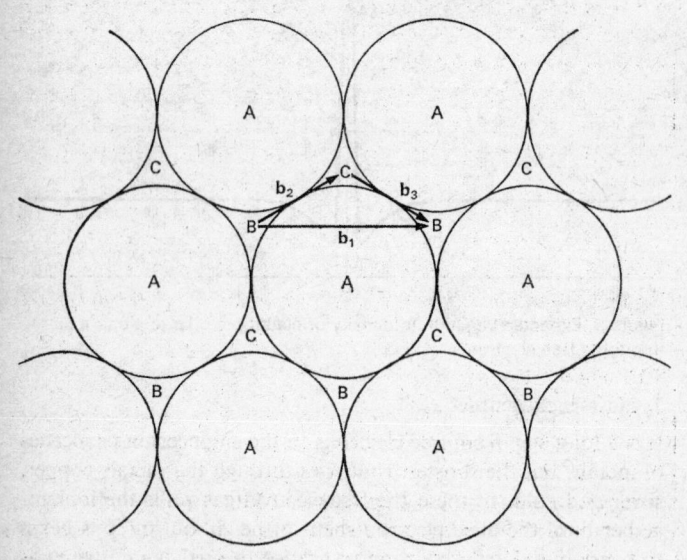

Figure 6 A close-packed plane

sequence ... ABABAB ... , and irregular sequences such as ... ABACBABCBAC ... , which all satisfy the same criterion of close packing.

Since copper, silver and gold do form FCC crystals there must exist in them something sensitive to the relative positions of more distant neighbours, e.g. electrostatic interactions of ions and electrons in different lattice cells. From our previous arguments we would expect these to be small effects, however, and this is confirmed by the ease with which these metals can be made to form 'annealing twins' and 'stacking faults' by various working and annealing treatments. An annealing twin consists of a reversal of the stacking sequence, as shown schematically in Figure 7, whereas a stacking fault may be regarded either as two such reversals, one immediately following the other, or as the result of slipping one half of the crystal relative to the other along a vector such as b_2 in Figure 6.

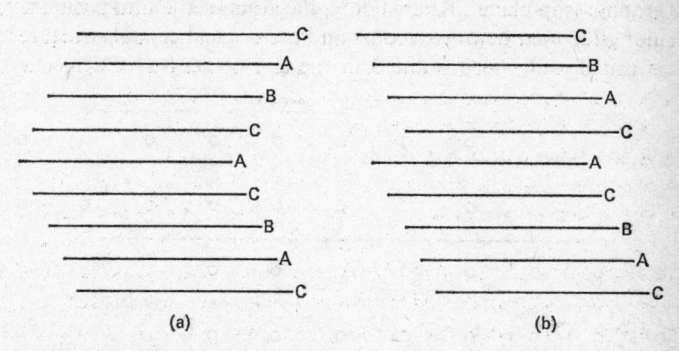

Figure 7 A twin (a) and a stacking fault (b) in an FCC sequence of close-packed planes

8 Strength and plasticity of metals

Metals are strong because they are soft. It is the glass bowl that breaks, not the silver one, when dropped on the floor, even though the silver can be easily scratched by the broken fragments. To resolve this paradox and understand mechanical properties it is necessary to distinguish between the inherent strength of matter and the overall strength of large pieces. A convenient measure for comparing strengths of different materials is the elastic *strain* at failure. When multiplied by the appropriate modulus of elasticity this gives the failure stress. Silicon is brittle at temperatures below 600 °C, but at this temperature its elastic failure strain is almost 1 per cent. Few metals and alloys withstand *elastic* strains as large as this before they break. The reason why brittle materials appear weak in practice is that ordinary samples usually contain small notches and other stress concentrators so that, locally, elastic strains of order 1 per cent are reached even while the overall applied load is still quite small. A soft, ductile metal, by contrast, usually begins to yield plastically at an elastic strain of order 0·1 per cent, and this smooths away the stress concentrations at notches before they become dangerously large.

Plastic deformation occurs by the slipping of crystal planes over each other, often by hundreds of atomic spacings, as shown rather simply in Figure 8. The slip takes place along a certain crystallographic 'slip direction' and usually on a crystallographic 'slip plane'. After sliding, the atoms settle into positions alongside their new neighbours and the original crystal structure is usually fully reconstituted in the slip plane. In FCC metals

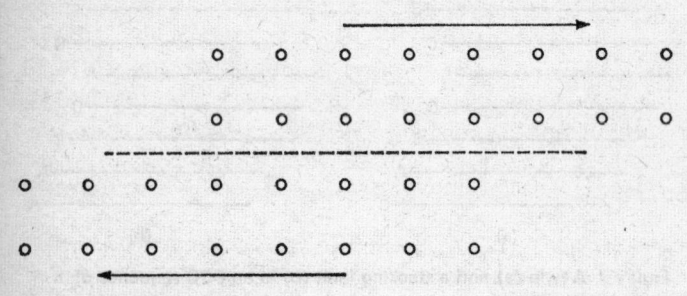

Figure 8 A simple slip process

slip usually follows the close-packed directions on the close-packed planes. The vector b_1 in Figure 6 represents in fact the smallest 'slip vector' that can reconstitute the original crystal structure. Close-packed directions and planes are preferred because, in a sense, they enable the sliding atoms to 'ride smoothly'.

The free-electron bond favours the slip process for two reasons. First, because it does not oppose the sliding of one atom over another, so long as they do not pull apart. There are no interatomic bonds to be broken since the atoms are bonded to the free electron-gas, not to each other. The close-packing requirement in metals such as copper, silver and gold is best satisfied during slip if the atoms move in a series of zigzag steps, along the vectors b_2 and b_3 in Figure 6 rather than along b_1. Slip in this case is closely related to the faulting process shown in Figure 7(b).

The second and more important effect of free electrons is in providing simple close-packed crystal structures which have smooth planes and directions for slip. It is this that makes the material soft and ductile. When the crystal structure is complex, as in Laves phases and related substances such as $CuAl_2$ and 'sigma' phases in certain alloy steels, the material is hard and brittle despite the fact that it is made from good metals and bonded by free electrons.

9 Dislocations

To follow the story further we must examine the detailed atomic movements that occur during slip. Crystals are elastically flexible objects and it takes time for an elastic disturbance to spread from one side of a crystal to another. It is thus virtually impossible for all the atoms on a slip plane to slide simultaneously. Rather, slip begins in one small area of the plane and then spreads radially outwards from it, over the rest of the plane, like a wave spreading across the surface of a pond. As shown in Figure 9, while this process is taking place the slip plane is divided into a slipped area, which is bounded by the slip front or 'dislocation line' as it is called, and an unslipped area outside this line.

Recent experiments have amply confirmed theoretical predictions about the occurrence of slip by the generation and propagation of dislocation lines. Slip occurs mostly by the propagation of 'unit' dislocations each of which produces one quantum of slip, e.g. equal to the vector $\mathbf{b}_1$ (cf. Figure 6) on a close-packed plane. It has also been proved that in most crystals, and certainly in engineering metals and alloys, natural imperfections of the structure act as good sources of dislocations and that the process of slip itself increases the dislocation content of the material. There is thus never any shortage of dislocations in such materials and the yield stress is essentially the stress required to make these dislocations glide across their slip planes despite various obstacles that stand in their way.

The first obstacle to consider is the periodic field of atomic forces in the crystal. Figure 10 shows the atomic structure in a section through one type of simple dislocation. Such structures have now actually been resolved in certain materials with the electron microscope. Atom A is in the half-slipped position, sitting vertically above B at the centre of the dislocation. We may suppose the atoms to the left of A to have slipped already, towards the left. To move the dislocation by one atomic spacing towards the right, atom C has to be pulled up vertically above D; atom A then slides down to the same position relative to B that E occupies at present relative to F, and correspondingly smaller adjustments are made in the positions of more distant atoms.

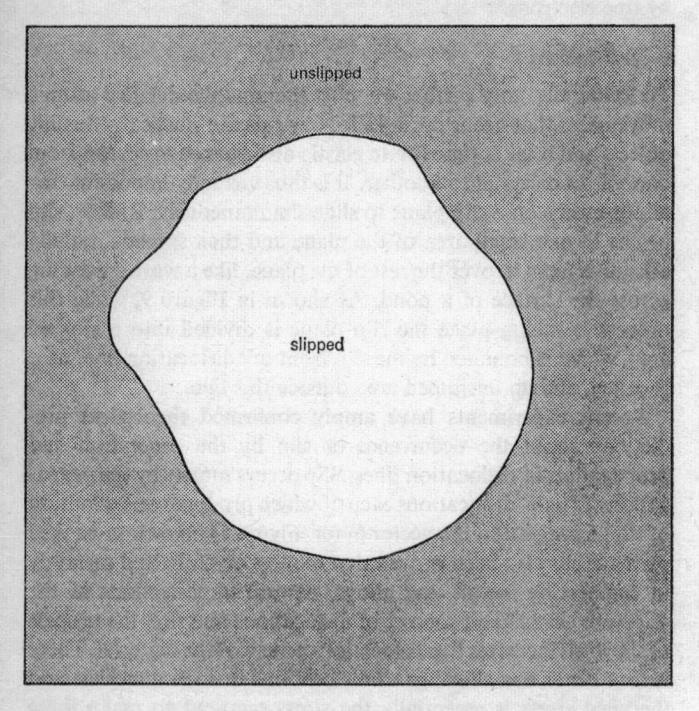

Figure 9 Plan view of a plane in process of slip, showing the slipped and unslipped areas, and the dislocation line which forms the boundary between them

In Figure 10 atoms C and E are symmetrically positioned about the centre of the dislocation. Atom E has almost completed its slip movement but, since it can fully complete it only if the dislocation glides away to the right, it exerts a force on the dislocation pushing it forward. But by symmetry, atom C exerts an equal and opposite force pushing the dislocation backward. The same holds for all other symmetrical pairs of atoms about the dislocation. In this position of the dislocation, therefore, the periodic crystal forces offer no resistance to its motion.

This is not true however for all positions of the dislocation. In transferring its centre from A to C the dislocation passes through unsymmetrical configurations in which the atomic forces no longer cancel in pairs. There is then a resistance to the movement of the dislocation, and it turns out that this is large when the dislocation is *narrow*, i.e. when the width of the belt of badly disarranged atoms in the slip plane, e.g. from E to C in Figure 10, is small. Narrow dislocations are expected when the atoms are

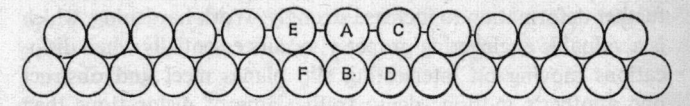

Figure 10 A simple dislocation

sensitive to their relative positions, and the hardness and brittleness of substances such as diamond and silicon is largely due to this effect.

The free-electron bond favours wide dislocations and a small lattice resistance to slip, since it is insensitive to the relative positions of the atoms. Ideal conditions for wide dislocations are found on close-packed planes in metals. Not only is the field of periodic atomic forces particularly smooth on such planes; the atoms are also provided with an excellent half-way position. In sliding according to the vector $\mathbf{b}_1$ in Figure 6, for example, the atoms may move first to C by the vector $\mathbf{b}_2$, and then complete their unit of slip, using the vector $\mathbf{b}_3$. This gives the dislocation the opportunity to spread out into a wide ribbon, rather than a narrow line, in the slip plane. The leading edge of this ribbon then produces a slip $\mathbf{b}_2$ into the faulted position, the centre of the ribbon consists of a stacking fault, and the trailing edge produces the remaining slip $\mathbf{b}_3$. In copper, silver and gold, the width of such dislocations is about ten atomic spacings. In a-brass (e.g. 70 per cent Cu, 30 per cent Zn), FCC stainless steel (e.g. 18 per cent Cr, 8 per cent Ni, in Fe), and several related alloys, even wider dislocations are obtained. In some cases the leading and trailing parts of the dislocation are so easily separable that large areas of stacking fault can be produced across the slip plane.

Such materials are intrinsically soft. Any hardness they possess is due to alloy and impurity atoms, foreign particles and precipitated phases, grain boundaries and lattice strains, which all serve as obstacles to the motion of dislocations. The more the material is purified and the structure annealed, the softer it becomes, and this softness persists even at the very lowest temperatures and under impact conditions of loading.

Such materials also work-harden strongly; i.e. the lattice strains produced as a result of plastic working cause the yield stress for further deformation to increase strongly. Work hardening, which is a valuable engineering property, occurs essentially when dislocations moving on intersecting slip planes meet and obstruct one another's motion; dense traffic jams of dislocations then build up at such places of intersection. This effect is particularly pronounced when the individual dislocations split apart to form wide stacking faults. Such a fault provides an almost impenetrable plane barrier to dislocations trying to glide through it on other planes. Very high hardnesses can be developed in this way without the material becoming excessively brittle.

The other main method of hardening such metals is through controlled precipitation of fine particles in the crystals by careful alloying and heat-treatment methods. Sometimes these particles are formed right on the dislocation lines and anchor them in position. Often they are dispersed throughout the entire crystal and provide a series of obstacles which the dislocation must overcome in its passage through the slip plane. One important class of creep-resistant alloys for jet engines consists in fact of FCC stainless steel hardened by fine precipitates of such things as, for example, niobium carbide.

The wideness of the dislocations in such materials is also important for their high-temperature strength. At temperatures high enough for solid-state diffusion to occur many more opportunities exist for dislocations to by-pass obstacles than at, say, room temperature. Most materials become soft under these conditions. This by-passing occurs mainly by movements of dislocations in directions *out* of their slip planes and such processes can occur only with difficulty when the dislocations exist as broad ribbons in their slip planes. It is hardly possible to move such a ribbon, at *any* temperature, in any direction other than in its own plane, and if the movement in this plane is obstructed by fine precipitates that persist at high temperatures, the material will remain strong even when hot. This is the basis of a large class of modern creep-resistant alloys. The starting material is an FCC metal or alloy of high melting point that forms wide dislocations. Further alloy additions are then made to it, some which may widen the dislocations further, e.g. cobalt, others which may form stable high-temperature precipitates, e.g. alloy carbides.

Part Four **Plasma Physics**

A plasma is essentially an ionized gas, that is a mixture of ions and electrons. The ionization may be partial or complete. The plasma has been called the fourth state of matter and in many ways has to be treated as such, its properties being quite different from those of solids, liquids and unionized gases.

I have included this work partly because it is a relatively new, exciting and rapidly developing branch of physics and partly because it holds out the prospect of achieving a controlled nuclear fusion process. In brief the distinction between nuclear fission and nuclear fusion is that in fission a heavy nucleus is split into fragments, with the release of large amounts of energy, and in fusion two light nuclei are joined together, again with release of energy. To understand why energy can be released in both these processes refer to Reading 5, Figure 4 on p. 88. This shows how the binding energy per particle varies with the mass of the nucleus. It indicates that this binding energy is increased either by splitting a heavy nucleus (e.g. ^{235}U) into nuclei of masses greater than 100; or by joining together light nuclei. An increase in the binding energy per particle means a release of energy when the change takes place.

The paper by M. F. Hoyaux is a review of some of the ways in which plasmas can be generated, an outline of their properties and an indication of some of the possible applications. Difficult mathematics is avoided and this paper provides a very good introduction to the subject.

12 M. F. Hoyaux

Plasma Physics and its Applications

M. F. Hoyaux, 'Plasma physics and its applications', *Contemporary Physics*, vol. 7, 1966, no. 4, pp. 241–60.

1 Introduction

Plasma physics is a rapidly expanding field of science. For quite a long time it coincided with the study of electrical discharges in gases; its many applications have appeared only recently. The conjunction of magnetohydrodynamics and plasma physics promises industrial applications such as the direct conversion of thermal energy into electrical energy without the use of rotating machines, as well as new rocket types which may have applications entirely different from those of the chemical rocket. In the earth's atmosphere fast-moving bodies create plasmas, and this is likely to afford new fields of interest both in aeronautics and in radio communications with manned missiles. Astrophysics is concerned with plasma not only in the interior of stars but also in cosmic space itself, where important magnetohydrodynamic phenomena are taking place.

Plasma physics now also meets what has been called the most formidable challenge ever faced by human knowledge: the *controlled thermonuclear process*. If and when this is successfully achieved, humanity will be provided with an almost costless source of energy for centuries to come.

Plasma physics has often in the past appeared to be unpredictable and unreliable. The repetition of an experiment under apparently identical conditions could sometimes lead to very different results. Most of the reasons for such discrepancies have now been explained. First of all, the use of oversimplified models for the plasma, rather directly inspired from the kinetic theory of gases, can be extremely misleading. Second, the degree of purity of the materials involved has to satisfy high requirements, unprecedented originally, but now commonly encountered; for instance in neutron physics and semiconductor applications. Plasma phenomena *in a gas* may be strongly dependent on the presence of impurities in very small quantities which, owing to conveniently placed excitation levels for instance, contribute in a spectacular manner to the onset of ionization. The phenomena *at the walls* of a containing vessel may be strongly dependent on monomolecular or at least very thin and almost uncontrollable adjacent layers. Even if the wall surface has been thoroughly cleaned before the start of an experiment, one can never be sure that some unpredictable chemical reaction has not created a new layer. It is not surprising then that much of the older information is unreliable and exhibits intolerable discrepancies. However, with a continual improvement in experimental conditions, better and better results are now becoming possible.

In the theoretical field also, and especially in the case of plasmas contained in significant magnetic fields, a tremendous contribution has been made during the last few years towards replacing the earlier oversimplified models. The behaviour of charge carriers in magnetic fields is very complicated, related in classical mechanics to gyroscope theory, but even more difficult. So we have a dilemma: for highly reliable results the calculations may become unmanageable, but if we make use of simplified theory the final results may be unreliable.

Finally, we would like to emphasize the tremendous development of diagnostic methods, mainly in connection with thermonuclear approaches. In a well-known paper presented at the Geneva Conference of 1958 not less than eighteen groups of diagnostic apparatus were itemized, and new ones have been discovered since then. In spite of difficulties in reproducing some, the total amount of experimental data is becoming quite large, and allows a good control both of the basic data and the theoretical models.

2 Definition and onset of the plasma state

There are several definitions of plasma, not all of which are in exact agreement. In principle, a plasma is an ionized gas, that is a gas in which at least a proportion of the atoms or molecules is broken down into charge carriers. Three main kinds of charge carrier will be considered in what follows: free electrons, positive ions and negative ions. The ions may carry a single or multiple charge; in addition, they may be atomic or molecular, or even more complex structures (for instance, an ordinary ion surrounded by a cloud of polarizable molecules).

Since an initial non-ionized gas is electrically neutral, we have in most cases an approximate balance between positive and negative charges (at least if sufficient volumes are concerned), and this neutrality is sometimes considered as the definition of a plasma. It can be shown theoretically and experimentally that, above a

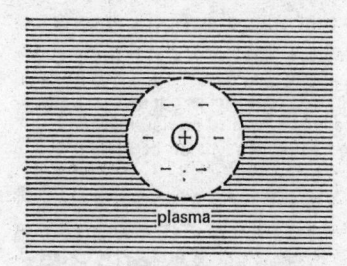

Figure 1 Debye screening length. When a 'test charge' of one sign is introduced into a plasma, it tends, owing to electrostatic attraction, to be surrounded by a cloud with an excess charge of the opposite sign; the field of the test charge is 'screened' and goes to zero more rapidly than Coulomb's law would indicate. The Coulomb potential is attenuated as exp $(-r/R_D)$, where R_D is the Debye screening length

certain density of charge carriers, the plasma has a very strong tendency to maintain its neutrality. An excess of charge carriers of one sign can only be maintained for very short distances (Debye screening length – see Figure 1) or during very short times (Plasma electrostatic oscillations – see Figure 2).

An important variable is the *degree of ionization*, which means the percentage of those gaseous atoms or molecules present which have been broken down into charge carriers. A gas may be weakly ionized, strongly ionized and even fully ionized. Some authors consider that a plasma is necessarily a fully ionized gas but historically the name 'plasma' was introduced (by Langmuir) well before much interest began to be devoted to fully ionized gases.

However, it is extremely convenient to think about 'the plasma state' as a fourth state of matter, in addition to the solid, liquid and gaseous states (Figure 3). Most of the properties of changes of state, heat of transformation, phase mixtures, etc., can be

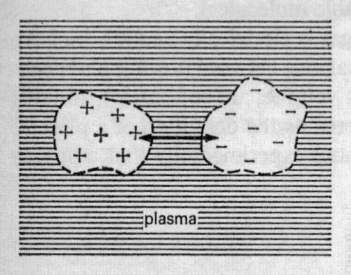

Figure 2 Plasma electrostatic oscillation. Local excess charges of both signs attract each other electrostatically and tend to be neutralized. However, if collisions are scarce, the first tendency is an oscillation of high (e.g. 10^4 MHz) frequency

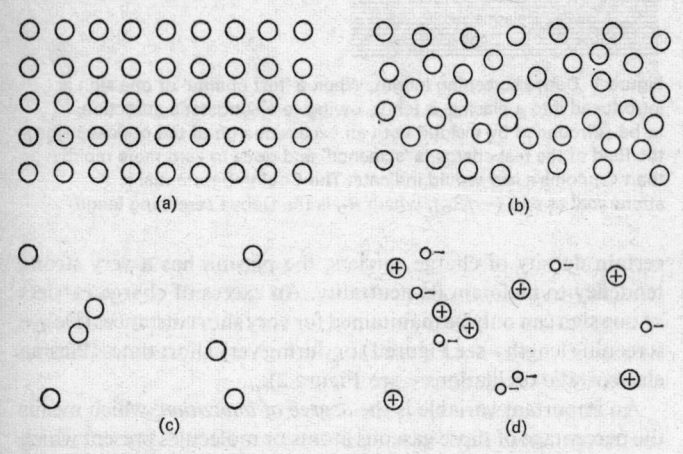

Figure 3 The four states of matter. (a) The solid (crystalline) state; (b) the liquid state; (c) the gaseous state; (d) the plasma (fully ionized) state. An incompletely ionized plasma can be regarded as a mixture of (c) and (d) states

extended without too much difficulty. However, two major differences do exist: first, the 'heat of transformation' developed on entering the plasma state is of about one decimal order of magnitude greater than for the others; second, the concept of phase mixtures with constituents at widely different temperatures is the rule rather than the exception in plasma physics, and there is no equivalent in the other states of matter.

One of the most conspicuous properties of a plasma is that it is, by contradistinction with ordinary gases, a good electric conductor. High-temperature fully ionized plasmas can exhibit an even lower resistivity than the best metals; in such cases the assumption

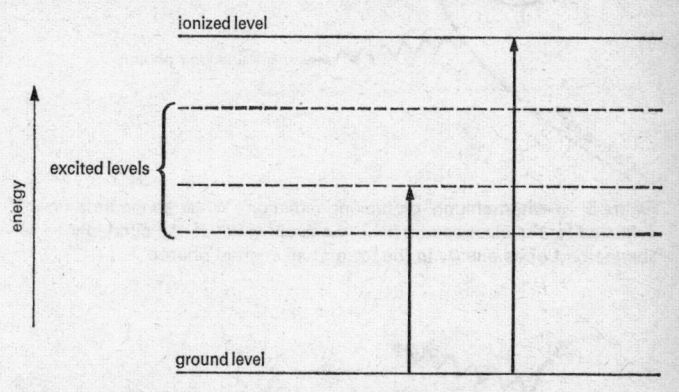

Figure 4 Conventional representation of energy levels. Excitation and ionization energies from the ground (i.e. normal neutral) state are plotted vertically. This figure explains why, in almost every plasma incompletely ionized, excitation appears as a significant by-product of ionization

of zero resistivity is, in general, an extremely valuable simplification. Of course, gases with a low degree of ionization are in general very far away from this ideal state.

Another property of a plasma is that at most times it emits significant amounts of light and other radiations. In low-ionized plasmas most of this radiation arises from atoms or molecules in an excited state, which are inevitably generated as a sort of by-product of the generation of charge carriers (Figure 4). In strongly or fully ionized plasmas the radiation is emitted by the free electrons. It is well known in field theory that an accelerated electric charge emits radiation in proportion to the square of the charge and the square of the acceleration. In a plasma the

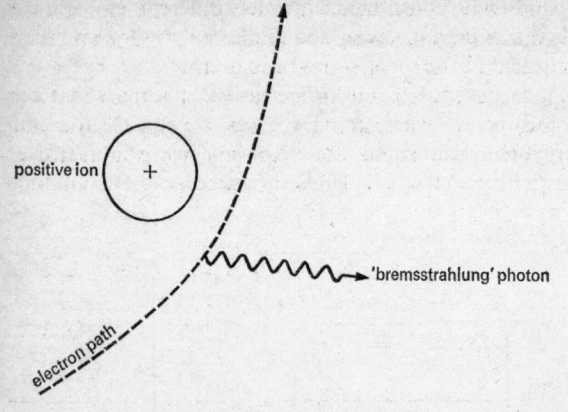

Figure 5 'Bremsstrahlung' or 'braking radiation'. When an electron is deflected by the electrostatic field of a positive ion, it can suddenly liberate part of its energy in the form of an emitted photon

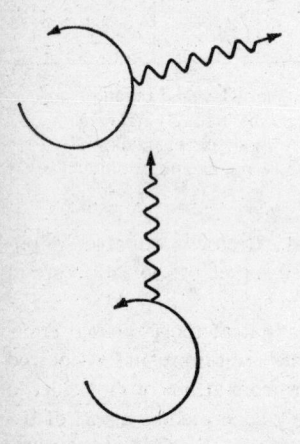

Figure 6 Cyclotron or synchrotron radiation. The same phenomenon can occur if the deflecting field is magnetic instead of electrostatic. This is termed 'cyclotron radiation' if the relativistic correction is negligible, and 'synchrotron radiation' if it is not

electrons are constantly deflected either by the electric micro-field of the positive ions, or by the general magnetic field if any. In the first case, physicists speak of 'bremsstrahlung'; in the second, of 'cyclotron-' or 'synchrotron-radiation' (Figures 5 and 6).

How can matter be brought, at least partially, into the plasma state? Like the change of state from solid to liquid and from liquid to gas this can be achieved by simple heating; however, in most cases other phenomena are involved. Generation of charge carriers (and we shall consider at the same time that of excited atoms or molecules, which in most cases inevitably accompanies it) can be achieved in two different ways: either by an action on the gas itself, or by an action on solid or liquid walls bounding the gas.

(i) *In the gas itself* ionization and excitation can be achieved by simple heating. Some gases, like caesium vapour, are extremely responsive to this method and can be used as 'seeding materials' in order to render other gases conductive at temperatures of 2000 °C and even less. But most gases require temperatures in the range of 5000 – 10 000 °C in order to exhibit any significant conductivity (Figure 7). The phenomena can be regarded in

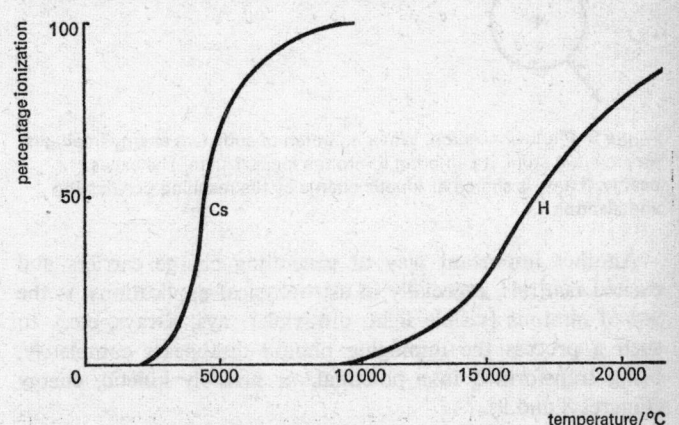

Figure 7 Thermal ionization. The percentage of ionized atoms at a given temperature, and especially at relatively low temperatures, differs widely from one species to another. Here is shown a comparison between caesium and hydrogen

different ways, either thermodynamically, as a special kind of chemical reaction obeying well-known relevant formulae, or kinetically, as the result of collisions of increasing violence as the temperature rises; both theories accord well with experimental results. For practical reasons, thermal ionization and excitation are generally confined to high-pressure gases (i.e. gases at pressures in excess of, say, one-tenth of an atmosphere).

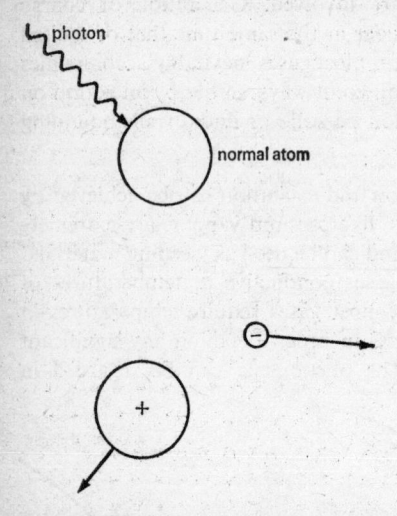

Figure 8 Photo-ionization. When a photon of sufficient energy impinges on a neutral atom, it can bring it into the ionized state. The excess energy, if any, is shared as kinetic energy by the resulting positive ion and electron

Another important way of generating charge carriers and excited neutrals, especially in astrophysical applications, is the use of photons (visible light, ultraviolet rays, X-rays, etc.). In such a process the impinging photon disappears completely, being transformed into potential, or possibly kinetic, energy (Figures 8 and 9).

In many types of electrical discharge, the electron component of the plasma plays the most active role in the generation processes.

It is well known that, when a fast electron collides with a

neutral atom or molecule, several types of phenomenon can occur. It may rebound 'elastically' (i.e. without change of kinetic energy) (Figure 10a) or induce an excitation (Figure 10b) or an ionizing phenomenon (Figure 10c) or, in the case of complex structures, induce a chemical decomposition (Figure 10d). In many types of plasma all these occur simultaneously. This is

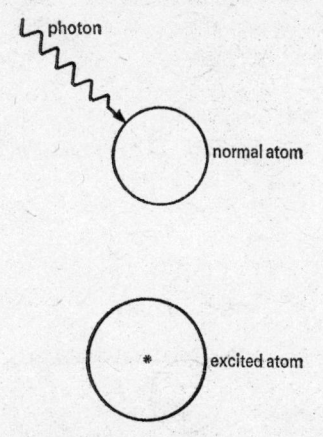

Figure 9 Photo-excitation. There is no opportunity for the excited atom to carry away a large amount of kinetic energy, at the same time satisfying momentum conservation. Thus photo-excitation is extremely energy-selective

especially characteristic of low-pressure gas discharges (i.e. below one-tenth of an atmosphere or so) because, at such pressures, the electrons are significantly accelerated between collisions by the electric field of the discharge, and the 'electron gas' tends to assume a higher temperature (kinetically defined) than the neutral component or the positive ions. The difference can be as high as two and even more decimal orders of magnitude.

Positive ions can, in theory, produce the same kind of phenomena, but, roughly speaking, the same order of *speed* has to be achieved, i.e. because of the much greater mass of the ion, an entirely different order of magnitude in *kinetic energy*. However, in any plasma phenomenon where fast positive ions are produced, this effect is likely to play a role.

Finally, we must mention a mode of carrier generation which, for instance in the moderate-pressure mercury arc, plays a significant role; it is the intervention of special excited states called 'metastable states' of the atom or the molecule. As far as

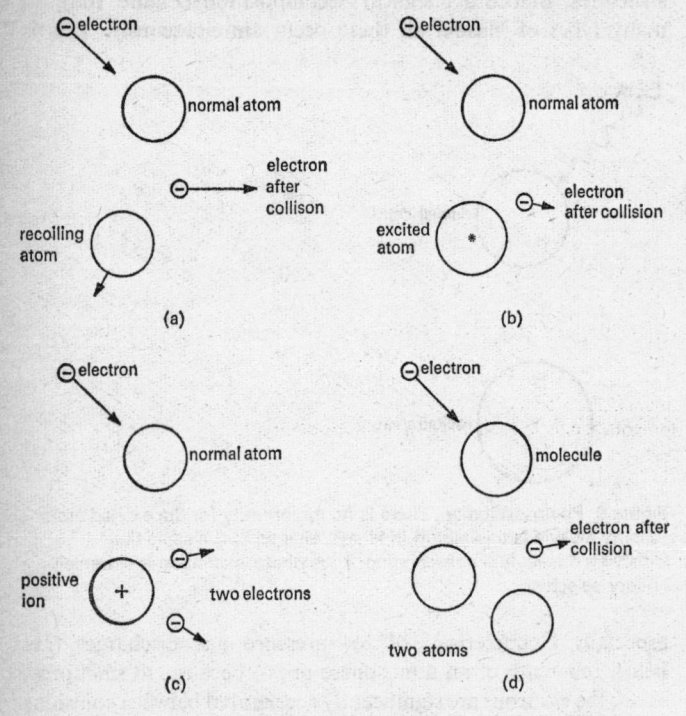

Figure 10 (a, b, c and d). Four types of collision between an electron and a neutral atom or molecule. The electron can rebound elastically (i.e. with the sum of the kinetic energies staying constant) or cause excitation or ionization, or chemical decomposition (e.g. decomposition of a diatomic molecule into two atoms)

generation is concerned, the metastable states arise in exactly the same way as ordinary states; but whilst the latter disappear in a time of the order of 10^{-8} s by photon emission, the metastable state cannot revert to normal in this way, since this would involve a transition which is 'forbidden' by quantum-mechanical selection rules. So an atom or a molecule in a metastable state of excitation can only release its energy through a collision. Several types of collision lead to ionization. The three main ones are:

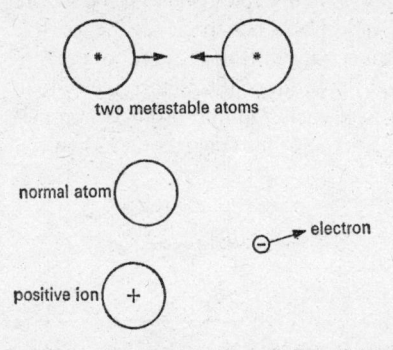

Figure 11 An example of a collision of the second kind. A collision of the second kind is one in which the total kinetic energy is increased (i.e. another form of energy is converted into kinetic). Here two metastable atoms of sufficient potential energy collide with each other; one of them is ionized and the other reverts to normal

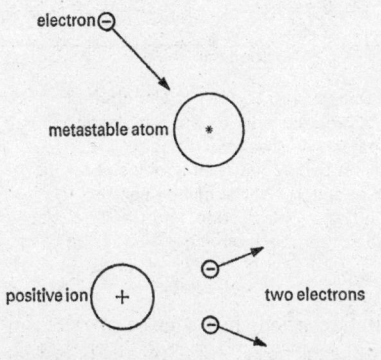

Figure 12 Step ionization. An atom already in a metastable state (resulting from a previous collision) is brought into the ionized state by a second collision

(a) two metastable atoms, each having more than half the energy required for ionization, collide with each other; one of them is ionized whilst the other one reverts to normal; (b) a metastable atom is struck by an electron or any projectile of sufficient kinetic energy and converted to the ionized state; this is designated as 'step-ionization', and in special conditions it can happen even with non-metastable states; (c) a metastable atom of some species A collides with a normal atom of another species B (hence, this is characteristic of gas mixtures only), the ionization potential of which is *below* the excitation potential of the metastable state in species A. The energy can be transferred and atom B becomes ionized, whereas atom A reverts to normal (Figures 11, 12 and 13).

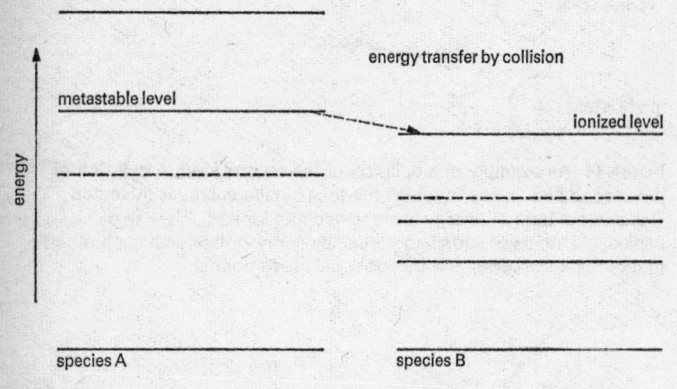

Figure 13 Ionization by energy transfer (cf. Figure 4). The atom of species A has been brought, by a previous collision, to a metastable state with a potential energy greater than that required to ionize an atom of species B. When a collision occurs between a metastable atom of species A and a normal atom of species B, the energy can be transferred. Atom B is brought into the ionized state, and the difference appears as kinetic energy, i.e. this is another example of a collision of the second kind

(ii) *At surfaces*. Let us now review briefly how charge carriers can be generated at solid or liquid surfaces. First of all, it is well known that mere heating of such materials releases electrons; in special conditions, positive and even negative ions can also be emitted. The impact of photons, electrons, positive ions, metastable neutrals and even neutrals in the normal state can lead to the emission of electrons if the conservation of energy is satisfied. In the case of metals, metastable atoms colliding with a surface can become ionized if the excitation potential of the metastable state is above the potential required for extracting an electron from

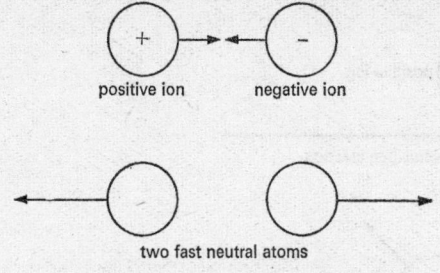

Figure 14 Direct recombination. A positive and a negative ion can recombine directly by mutual annihilation of their charges: two fast neutrals appear, with the overall centre of mass essentially at rest; this satisfies both energy and momentum conservation

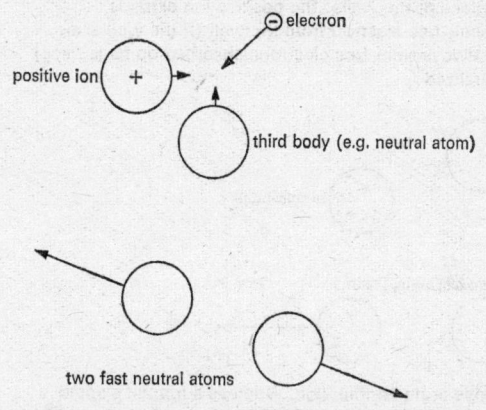

Figure 15 Third-body recombination. Recombination of a positive ion and an electron results in only one neutral particle, which would have to be fast in order to satisfy energy conservation, and slow in order to satisfy momentum conservation. So, direct recombination is generally impossible, and the intervention of a 'third body' (e.g. a neutral atom) is essential

the metal; in this case, the metal surface loses an electron to the metastable atom. In extreme cases, for example with caesium, this can happen even with a normal neutral.

In any plasma-physics phenomenon there is in general a delicate balance between charge-carrier generation and annihilation; the annihilation processes are much less numerous than

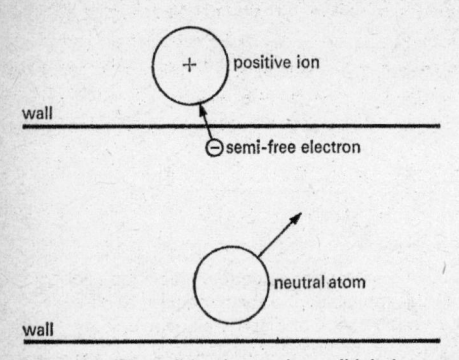

Figure 16 'Recombination at the wall.' At low pressures all the carrier annihilations take place at the walls: the positive ion extracts electrostatically a semi-free electron from the wall (if the wall is an insulator there are always semi-free electrons absorbed on its surface) and rebounds neutralized

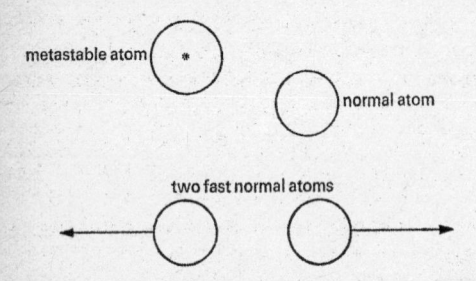

Figure 17 Metastable atom annihilation. Whereas a normal excited atom returns spontaneously to ground level by photon emission with a delay generally not exceeding some 10^{-8} s, a metastable atom can do so only by colliding with another body or a wall. For instance, here a metastable atom collides with a normal atom; it goes back to normal; the potential energy of the metastable atom is transformed into kinetic energy and shared about equally between the two atoms

those of generation. Annihilation can take place (by pairs) in the gas itself, with or without the intervention of a third body (Figures 14 and 15) or at the walls (Figure 16). Excited atoms or molecules can release their potential energy in collisions (collisions of the second kind) (Figure 17) or by striking the walls, but this is rather exceptional except with metastable atoms or molecules. In general, the energy is emitted in the form of a photon, in a spontaneous transition delayed only by some 10^{-8} s. Since the wavelength is characteristic of the gas used, it happens quite often that the radiation remains 'trapped' for some time, i.e. is re-absorbed

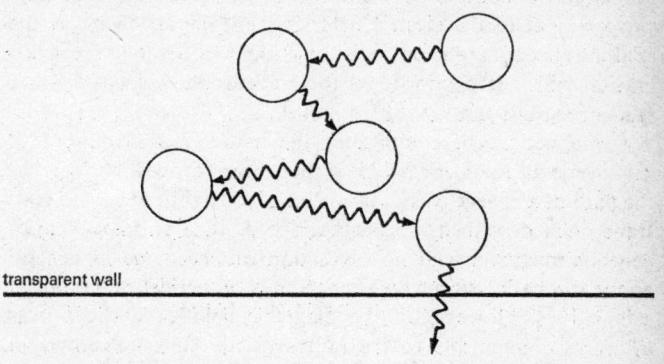

Figure 18 Radiation trapping. Since the radiation emitted by the plasma-excited atoms corresponds to a characteristic wavelength of the relevant gas, the probability of its being recaptured is far from negligible. Sometimes the same quantum of energy is re-emitted and recaptured more than a hundred times before finding a way out through a (supposedly transparent) wall. This is termed 'radiation trapping' and obeys the ordinary law of diffusion

and re-emitted by several neutrals in sequence before finding its way out (Figure 18).

In fully ionized plasma the temperature is in theory such that no recombination can take place; thus the balance between generation and annihilation is irrelevant in the steady state, and the equations are considerably simplified.

3 Motion of charge carriers

The macroscopic properties of the plasma state depend very strongly upon the laws governing the motion of charge carriers in electric, magnetic and (in astrophysics) gravitational fields. Motion in a gravitational field alone or an electric field alone is extremely simple. In the first case the plasma simply falls freely, unless there is an orbit under some kind of central force as in the motion of a planet around the sun. In the second case there is of course a tendency to charge separation, the positive carriers drifting one way and the negative ones the other. This corresponds to the passage of an electric current or, if the latter is hampered in some way, to a polarization of the plasma. An individual carrier should move with a uniform acceleration directly proportional to the electric field; but in general, owing to the braking effect of collisions, it is the velocity rather than the acceleration which is proportional to the field; the velocity in unit field-strength is termed the 'mobility'.

A magnetic field complicates the matter significantly. If a homogeneous magnetic field is acting alone, it is well known that the path of a charge carrier is a helix of constant radius and constant pitch described at constant speed. In a strongly heterogeneous magnetic field no prediction can be made in general about the path, except by very tedious numerical or graphical calculations. However, if the magnetic field varies little over distances comparable to the radius of the helix, or slowly in comparison with the time it takes the particle to make one turn, an extremely interesting approximate method, called the orbiting centre method, can be applied. According to this method, the path of the charge carrier is considered as a circle of variable radius and mobile centre, and, as a first approximation, the motion of the instantaneous centre is considered instead of that of the particle. The forces responsible for this motion are calculated by regarding the instantaneous circular orbit as a loop carrying a current equal to the charge of the carrier, divided by the duration of one revolution. In the calculation of such a force, a very important parameter is the so-called 'magnetic moment' of the loop, which is equal to the current times the area enclosed by the loop. If the magnetic field varies slowly, either as the result of a true time variation, or as the result of a drift of the instantaneous centre, the magnetic moment remains invariant. In the case of more rapid variations the situation is less clear, and there are definitely cases for which it is not invariant, although in general the tendency to invariance has been shown to be greater than elementary theories would lead one to expect.

The concept of magnetic moment enables one to introduce a new form of potential energy, termed 'magnetic potential energy', equal to the scalar product of the magnetic induction and of the magnetic moment (defined vectorially with a direction perpendicular to the plane of the orbit). The most stable configurations are those for which this potential energy is minimum. Hence, two important conclusions: (a) the 'arrow' of the magnetic moment tends to point in the opposite direction to the magnetic field, and (b) orbiting particles tend to avoid the regions of strong magnetic field and to drift toward the weaker parts.

Point (a) is related to the concept of 'plasma diamagnetism': the reaction of the plasma is always a tendency to weaken any pre-existing magnetic field, and even, in extreme cases, to cancel it. Point (b) is related to the concept of 'magnetic pressure' about which more will be said later.

If the magnetic moment is invariant, the magnetic flux embraced by the instantaneous orbit is invariant too. This is related to a third important concept, that of 'frozen magnetic lines of force' which will be explained later.

When the magnetic field is acting in conjunction with any other kind of force, either as a result of an electric or gravitational field, or as the result of a gradient in the magnetic potential energy defined above, or of an inertial (e.g. centrifugal) force resulting from any motion other than the instantaneous orbiting motion, paradoxical situations arise which are related to the theory of the gyroscope.

A first guess would be that the particle subjected to the influence of a magnetic field and another force will spin in the magnetic field with an instantaneous centre of motion drifting in the direction of this other force. This is entirely false. If a gyroscope is spinning freely and we apply a force, its tendency will be to drift, not in the direction of the force, but in the direction perpendicular to it. Qualitatively, the situation is exactly the same for orbiting charge carriers. After allowing for that component of the force parallel to the magnetic lines of force (which makes the centre of motion drift along this line, unhampered) the motion is always perpendicular to both the magnetic field and the other force.

If we go into further details the situation is even more paradoxical since, if the additional force arises from an electric field, positive and negative charge carriers tend to drift in the same direction with the same speed, whereas for non-electric forces

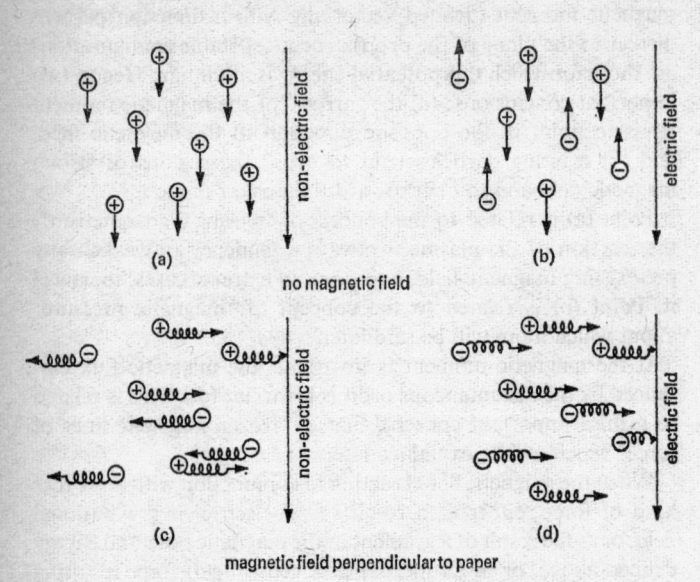

Figure 19 Paradoxical motion of charge carriers in a strong magnetic field. (a) A non-electric field-force acting alone: the plasma as a whole 'falls' freely under the action of this force; there is no charge separation, and therefore no current or polarization in the plasma. (b) An electric field acting alone: positive and negative charge carriers drift in opposite directions; there is an electric current through the plasma, or, if the current is hampered, at least a polarization. Case (c) is identical to case (a), but with a strong magnetic field perpendicular to the paper. Not only is the motion of charge carriers perpendicular to the non-electric field instead of parallel, but there is also a charge separation, hence a current or a polarization. By contradistinction, in case (d), where the non-electric force is replaced by an electric one, all particles drift together without any tendency to charge separation

there is always a charge separation and a tendency for a current to flow, or at least for the plasma to polarize if the flow is hampered for some reason. Hence, at least in the absence of collisions, *the electric field is the only one which is unable to cause a current across the magnetic lines of force!* This is true only for a static field; if the field is slowly varying, the drift velocity is varying too, so that there is an inertial force, *and this inertial force* (not the electric field!) can be considered as responsible for the current or the polarization.

The different parts of Figure 19 illustrate those paradoxes.

The above discussion is strictly true only in the absence of collisions, i.e. in a perfect vacuum, but actually particles interact with each other and with the neutral gas if any. This complicates the situation significantly. An extremely good simplifying model may be constructed by assuming that, by and large, the effect of the collisions is equivalent to a viscous drag. One finds that the key parameter is of the form B/p_0, where B is the magnetic induction and p_0 the so-called 'reduced pressure', i.e. a parameter proportional to the density of neutrals. If this parameter is large, the motion is essentially the same as in a vacuum; if it is small, the motion is essentially the same as in the absence of a magnetic field, i.e. the drifts are parallel to the forces and proportional to them. In intermediate situations the drift is neither parallel nor perpendicular to the force, and elaborate considerations of tensor algebra are needed. We want to emphasize the fact that, if p_0 is large, even strong magnetic fields can have little influence on the motion of the charge carriers.

4 Plasma confinement by magnetic fields

One of the most challenging questions in plasma physics is whether or not it is possible to confine a plasma, preferably a plasma at a very high temperature, away from any kind of solid wall by interposing a so-called magnetic bottle. This question is important, not only in view of the possible thermonuclear reactor but also for the plasma magnetohydrodynamic generator and the plasma rocket, where the situation is less critical but where magnetic confinement, if successfully achieved, would lead to a significant improvement of efficiency.

According to the statements of the preceding section, if the collision rate is kept sufficiently low, the drift resulting from any kind of force superimposed on a magnetic field and perpendicular to it is always perpendicular both to the magnetic lines of force and to this extra force. This can be applied, in particular, to the force resulting from a pressure gradient in a plasma. If there is a pressure gradient perpendicular to the magnetic lines of force it generates a drift perpendicular to both, and since the pressure gradient is a non-electric force, the second paradox comes into play: it generates a drifting current. Still in the absence of collisions, there will be strictly no drift parallel to the extra force, i.e. to the pressure gradient. Another way to state this is that the interaction of the drifting current with the magnetic field balances exactly the pressure gradient. Still another way, correct only if the curvature of the magnetic lines of force is not too significant in relation to the other parameters involved, is the introduction of the concept of 'magnetic pressure'. It is well known that pressure and energy-density can be expressed in the same units, and thus are in general related to each other. If p is the plasma pressure (not to be confused with p_0 previously introduced), it can be shown that, when the curvature of the magnetic lines of force is not too significant, the following relation holds:

$$p + \frac{B^2}{2\mu} \simeq \text{constant.}$$

μ is the magnetic permeability, hence $B^2/2\mu$ is the density of magnetic energy. The linear pinch (Figure 20) is one of the simplest applications of this concept: when the current is increased, the magnetic field outside the ionized channel increases, too, and the channel tends to be more and more constricted by its own magnetic field.

Several configurations exist for which such a balance is approximately achieved; they have been popularized in articles about thermonuclear approaches and will not be described here. Elementary treatments of such confinement schemes fail to emphasize the major problem, which is one of *stability*.

The concept of stability in plasma physics is an extension of that in elementary mechanics. A pencil resting vertically on its point is in equilibrium between its weight and the reaction of the table; but it is not a stable equilibrium because any displacement from its equilibrium position tends to be amplified until the pencil falls on the table. In more complicated mechanics problems it can very well happen that some displacements have a tendency to amplify themselves, whereas others have a tendency to reduce.

In order that the equilibrium can be termed 'stable', it is necessary that *all* the possible displacements tend to be self-reducing. This might lead to a very complicated calculation, but fortunately mechanics provides us with theorems allowing a selection to be made. All the displacements will be self-reducing if every displacement in a finite sequence, the so-called principal displacements, is self-reducing.

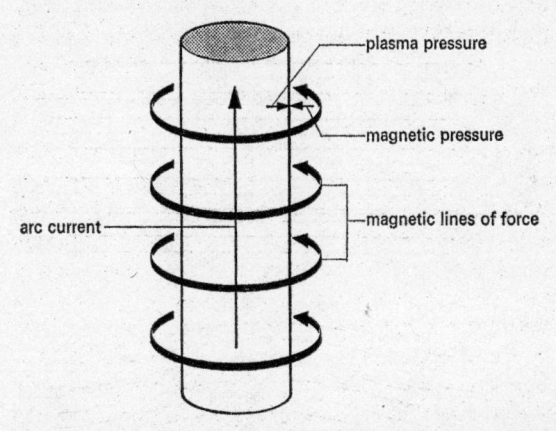

Figure 20 Magnetic confinement: the linear pinch. A plasma column carries an electric current; in certain conditions, an equilibrium can be attained between the plasma pressure outwards and the 'magnetic pressure' inwards. Unfortunately this equilibrium proves to be very unstable

The situation is the same in a plasma, with the difference, however, that the number of the so-called principal motions is enormously greater than in ordinary mechanics problems. Nevertheless a selection can be made so as to render the problem manageable.

In order to check the stability of a certain configuration of confined plasma, we have to imagine a deformation (the choice of a particular type of deformation is a matter of the calculator's skill) and to check if the plasma moves from this deformed configuration towards or away from the initial configuration. In many cases a sufficient condition of stability is that the deformed condition involves more energy than the initial one; then the situation is comparable to that of a pendulum displaced from its equilibrium position. Of course the energy here is rather complicated since it involves the thermal energy of the plasma, the magnetic energy, and, in astrophysical applications, gravitational energy.

In such calculations advantage can frequently be taken of the fact that, if the plasma electrical conductivity is sufficiently large,

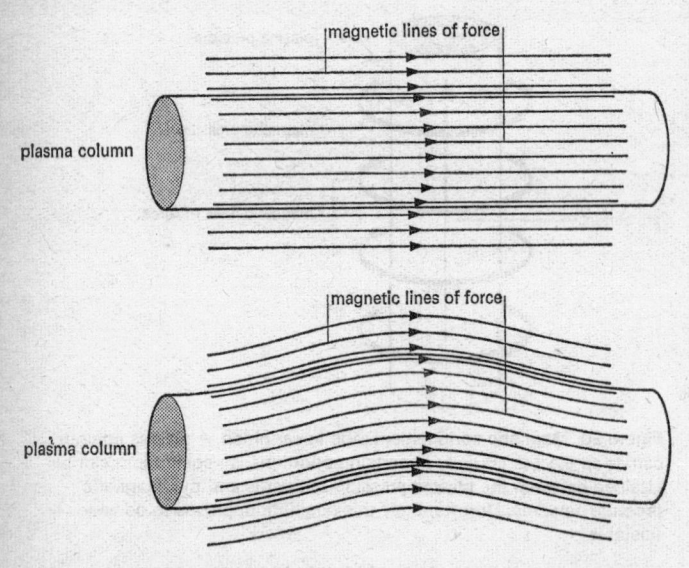

Figure 21 Frozen magnetic lines of force. There are many circumstances in plasma physics where the magnetic lines of force can be visualized as carried by the fluid and travelling with it. This concept lacks a strong scientific basis, but is extremely useful in practice

the magnetic lines of force can be considered as 'frozen' into it, an idea introduced by Alfvén. This has sometimes been bitterly criticized, because there is no accepted single definition of the notion of a magnetic line of force. However, at least in approximate arguments, the concept is very useful. In imposing a deformation on our initial plasma configuration we must remember that, at least as a first approximation, the magnetic lines of force are carried by the plasma (Figure 21). This is consistent with the remark in the preceding section, that the magnetic flux embraced by a slowly variable orbit is an invariant of the motion. When the orbit is displaced, not only does it embrace the same *number* of lines of force, but it can be considered that they are *the same ones*.

However, there is some contradiction between those two concepts and that of magnetic confinement. According to the latter the tendency is exactly opposite: there is less magnetic

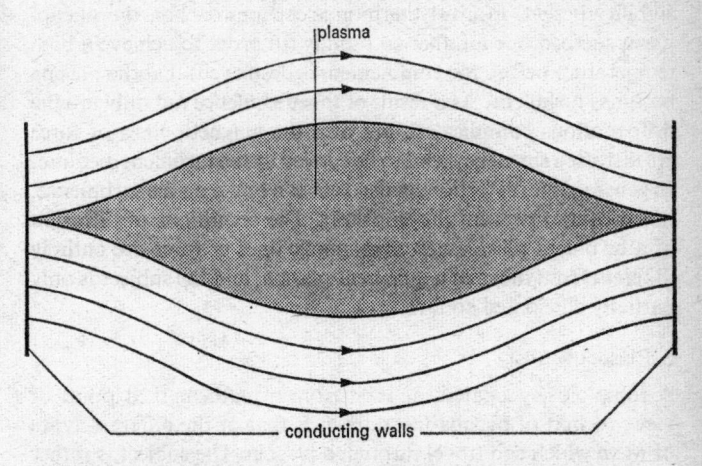

Figure 22 Magnetic lines of force 'attached' to the wall. This concept of 'frozen' magnetic lines of force can be applied to conducting walls, too. In this example the stability of a magnetic confinement is improved because the magnetic lines of force can be visualized as 'strings' attached to conducting walls

flux where there are more particles and vice versa. Let us consider a homogeneous plasma in a homogeneous magnetic field. For some reason the plasma density increases somewhere. The concept of frozen magnetic lines of force would lead to a local *increase* in the magnetic field, whereas that of magnetic pressure would lead to a local *decrease*. This contradiction is one of the reasons why instabilities occur. There are many entirely different types of instability, related for instance to charge separation under the combined action of a magnetic field and a non-electric force; in fact, even a simple enumeration of the different types of instability would be outside the scope of this paper.

The concept of frozen magnetic lines of force is applicable not only to plasmas, but also to conducting walls. In certain conditions the magnetic lines of force can be visualized as *strings*, and if one succeeds in attaching such strings to the walls the stability conditions are improved (Figure 22). It is also interesting to create situations in which the 'strings' are 'interwoven'.

A point which complicates significantly the subject of plasma stability is that, in most thermonuclear approaches, the phenomena succeed one another so rapidly (in order to achieve a high temperature before the confinement is destroyed) that the plasma becomes turbulent. As a result of this turbulence not only are the drift motions complicated, but also the magnetic lines of force are distorted since they tend to be carried by the turbulent medium. In general, this distortion tends to act as a brake on the turbulence, but it cannot prevent it completely. The conditions of diffusion of a turbulent plasma across magnetic lines of force are entirely different from those of a quiescent plasma, and the subject is only partially elucidated so far.

5 Plasma waves

A topic closely related, at least from a mathematical point of view, to that of plasma instabilities is that of the different types of wave which can travel through a plasma. The subject is rather complex, especially if there is a strong magnetic field throughout the plasma. Plasma waves can be roughly divided into three classes. In the first class we find the so-called 'electrostatic waves', characterized by the fact that there is a significant charge separation as a result of particle motions in the wave (longitudinal electric field). The wave can be depicted as a succession of layers alternately positive and negative. There are two such waves: the electro-acoustic wave, in which the motion of the electrons is predominant (high frequency), and the ionic wave, in which the motion of the ions is predominant (low frequency).

The second class of wave involves no charge separation, but a significant motion of the plasma, positive and negative particles travelling together. Again there are two principal types of wave: the first type travels *along* the magnetic lines of force; it can be regarded as an oscillation of the magnetic lines of force considered as strings carrying the plasma with them (frozen magnetic lines of force); this is termed an Alfvén wave. The second type, a magneto-acoustic wave, travels *across* the magnetic lines of force; it can be regarded as a special kind of sound wave in which, again in agreement with the concept of frozen magnetic lines of force, the magnetic pressure varies together with the ordinary pressure, and adds its effect to it, the speed being calculated from the usual formula for sound $(\partial p/\partial \rho)^{1/2}$, ρ being the density of the plasma. Both types are low-frequency waves; at high frequencies, 'freezing in' the magnetic lines of force tends to be destroyed.

The third class of wave involves neither charge separation nor significant drift motion, but comprises ordinary electromagnetic waves. As a first approximation, the influence of the plasma is equivalent to that of a refractive index. However, in strong magnetic fields the refractive index becomes a tensor; and phenomena analogous to, but not identical with, those of crystalline optics appear: bi-refringence, rotation of the polarization plane, etc.

In fact, the clear-cut distinction made above is somewhat misleading because it is only valid for motion parallel or perpendicular to the magnetic field. If the motion is oblique, the waves observed appear as mixtures of the preceding types, and the situation becomes extremely complicated.

A good knowledge of the properties of wave propagation through plasmas has proved of great value in sounding the ionosphere and as a diagnostic tool for laboratory experiments. Further, since a space capsule re-entering the atmosphere is, during certain phases of its descent, surrounded by a plasma created by its mere motion, any transmission of information to and from the capsule during this phase demands a good knowledge of the problem.

The waves considered so far are of small amplitude, i.e. the maximum *variation* of a parameter such as, for example, an electron density, during one cycle is very small compared with the average value of the same parameter. Such waves can be expressed by linear equations. If the amplitude becomes larger, the equations are no longer linear. Physically, what happens can be explained as follows: the wave velocity at a crest is greater than in a trough; as a result, each crest tends to come nearer and nearer to the preceding trough, until the front becomes very steep; at this stage the front becomes an effect in its own right, independent of the other parts of the wave (Figure 23). Such a front is called a shock wave.

Ordinary sound waves of large amplitude degenerate easily into shock waves, which are commonly observed when explosions

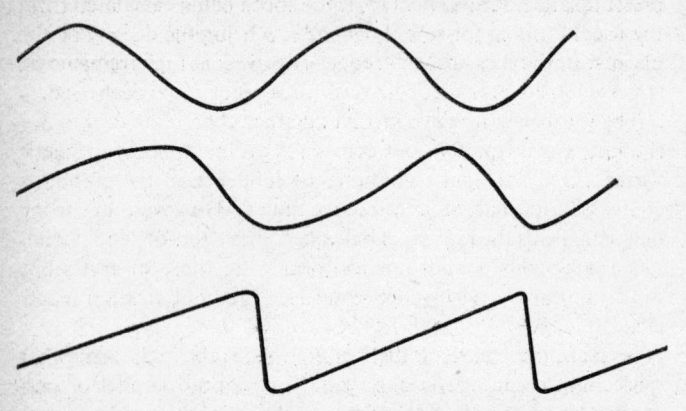

Figure 23 Degeneracy of a strong sound wave into a shock wave. In an ordinary sound wave the increase of pressure between a trough and a crest is negligible; but in a strong sound wave it is not. Since the velocity of sound increases with gas temperature, and the latter with pressure according to the law of adiabatic compression, a crest tends to travel faster than a trough, and the front becomes steeper and steeper until it is only a few mean free paths thick; then, parasitic effects (heat conductivity, viscosity, etc.) prevent any further steepening. The front then has a structure of its own, independent of the way it has been obtained, and is termed a shock wave. Several types of plasma wave can also degenerate into shock waves, the only requirement being a higher speed at the crest than at the trough

occur, when an aircraft crosses the sound barrier, when a diaphragm between two regions of unequal pressure is suddenly opened, or when a piston moves in a cylinder at a velocity greater than that of sound. In plasma physics several types of wave are liable to degenerate into shock waves if the amplitude is sufficient; among them are the Alfvén wave, the magnetosonic wave and their 'relatives' in the case of oblique flow. The magnetosonic shock wave has received most attention; it closely resembles an ordinary sound shock wave, with the additional feature that, the magnetic lines of force being frozen in the plasma, the passage of the shock front is accompanied by an increase of the magnetic field proportional to the density. Such an increase requires, in the transition layer, the presence of an electric current perpendicular to the direction of flow (Figure 24).

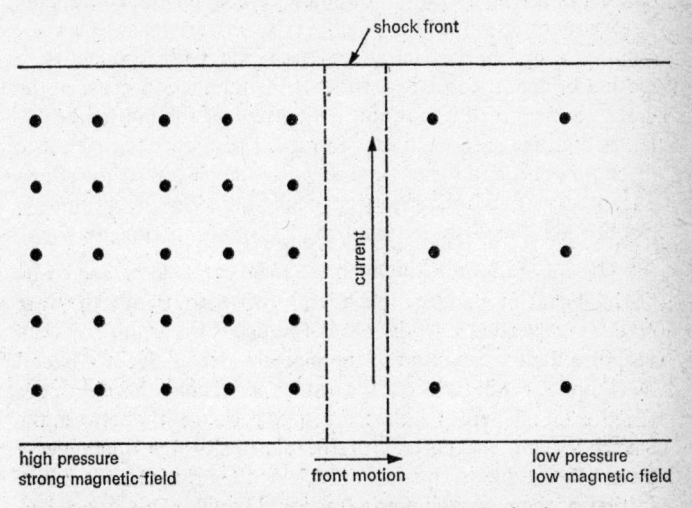

Figure 24 The simplest type of magnetohydrodynamic shock wave. This type of shock results from the degeneration of a magnetosonic wave. The laws are identical to those of a sound shock wave, provided the gas pressure is replaced by the sum of the gas pressure and the pressure of the (frozen) magnetic field

6 Plasma motors and generators

During recent years much attention has been devoted to the possibility of using a plasma as the working fluid in both motors (in fact, rocket or rocket-like motors) and generators.

We shall consider only the acceleration of a plasma as such, not the separate accelerations of the positive and negative particles which compose it. Two types of force can contribute to the acceleration of a plasma as a whole: a pressure gradient, and the interaction of a current with a magnetic field perpendicular to it. The pressure gradient is exploited in the ordinary nozzle and is not confined to plasma physics; any novel feature of a plasma rocket results from the exploitation of the magnetic force. The devices considered for such exploitation can be divided into three classes:

(a) The series injector, most easily understood in the form of parallel rails (Figure 25a), though the coaxial form will probably be eventually preferred for practical reasons. In this injector one and the same current circulates along the rails and across a plasma bridge. It can be shown that the interaction between the magnetic field of the rails and the current of the plasma bridge tends to propel the latter along the rails. The injection is in essence discontinuous, and the cooling down of the plasma at the points of contact with the rails is far from negligible. However, promising results have already been obtained by several experimenters.

(b) The shunt injector, in which the magnetic field is created by an independent pair of pole pieces (whether or not they are actually in parallel with the plasma bridge is irrelevant). We can visualize the cross-section of the pipe as a rectangle; two lateral faces are the pole pieces, the other two being the electrodes (Figure 25b). There is a plasma bridge permanently between the electrodes and the current of the plasma bridge undergoes a lateral thrust due to the magnetic field; the principle is identical to that of the magnetic pump for liquid metals. This device has, compared with the previous one, the advantage of being operated in a steady state, but the cooling by the electrodes is again a major shortcoming.

(c) The magnetic piston, in which a combination of coils around a cylindrical pipe creates a magnetic field with travelling maxima and minima. As we have seen before, the particles tend to avoid the maxima which act as magnetic pistons and sweep them down the pipe. The plasma is injected in bursts, but since several plasma blobs can travel down the tube at the same time, macroscopically speaking, the thrust can be rendered continuous. The major advantage is that the plasma blobs are, at least to some extent, confined by the magnetic field, and cool down much more slowly (Figure 25c).

In principle, any motor can be reversed in order to operate as a generator of electricity. However, the series injector does not seem

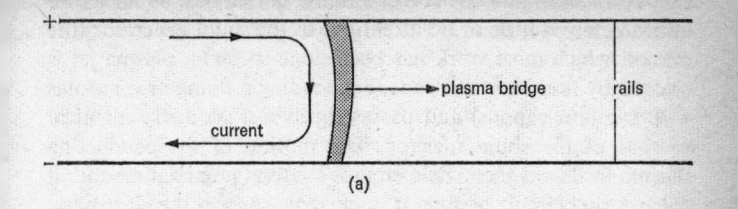

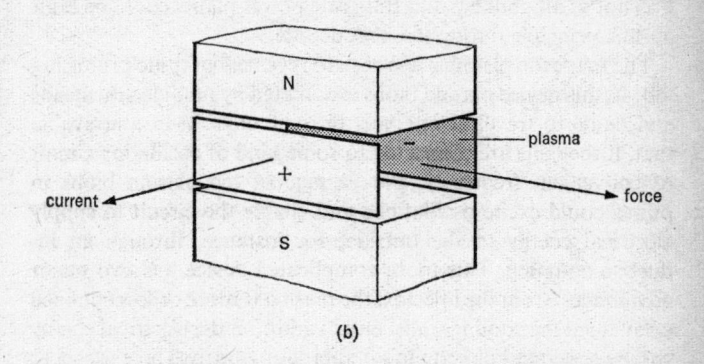

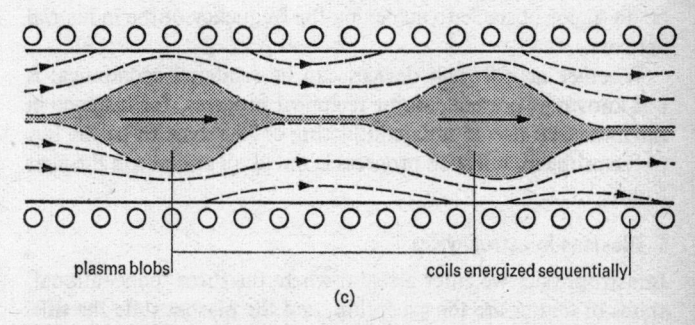

Figure 25 The three main types of plasma injectors: (a) series injector, in which the plasma bridge is accelerated by the interference of its own current with the magnetic field of the two rails; (b) shunt injectors, in which it reacts with a magnetic field created independently, and (c) magnetic piston, based on a special form of time-dependent magnetic confinement

easy to reverse, and the corresponding device has, as far as we know, received little or no attention. In the shunt generator (the one on which most work has been done so far) a plasma jet is created by non-electric means (e.g. seeding a flame or a hot gas with caesium vapour) and passes through a geometry identical to that of the shunt injector. The motion of the conducting plasma in the magnetic field creates a lateral polarization and, if the external circuit permits it, a current between the electrodes. It is not at all unlikely that full-scale power plants could be built on this principle during the next decade.

The magnetic piston is less easy to reverse, but quite promising too. In this device plasma blobs are created by non-electric means and made to travel at intervals through coils. It is conceivable that, if the coils are connected to some kind of oscillating circuit of convenient frequency, the passage of the plasma blobs in pulses could excite oscillations and enable the circuit to supply electrical energy to the outside; for instance, through an inductive coupling. This more complicated device has two major advantages: as in the injector, the plasma is more or less confined away from the cooling walls, and in addition the electrical energy can be generated directly in an alternating form. There seems to be no major obstacle to achieving the frequency of the industrial network.

Whether or not such devices can be rendered economical is not known at present. Major technical progress, for instance in the industrial use of superconducting coils, seems to be the key to the problem, but such progress is not at all unlikely in the near future.

7 Plasmas in astrophysics

In astrophysics we enter a realm where the three 'conventional' states of matter are the exception, and the plasma state the rule. Moreover, significant magnetic fields are frequently trapped in astrophysical plasmas, and the conditions for frozen magnetic lines of force are quite generally met.

Take for instance the sun, which for many purposes can be considered as a typical star. Practically all the matter is in the ionized, and even fully stripped, state (this expression describes matter in which the atoms have lost not just one but all their electrons). Already in the normally visible part of the sun, the so-called photosphere, the manifestations of plasma physics are many. The well-known sunspots are characterized by trapped magnetic fields in the thousands-of-gauss range. Their eleven-year periodicity is probably related to the motion of Alfvén waves or some related phenomenon along the lines of force of a general dipole field. Especially in spotted areas, a substantial fraction of the emitted energy seems to be in the form of plasma waves or shock waves. If we go towards the normally invisible 'atmospheres' of the sun, the chromosphere and the corona, the plasma-physics phenomena are perhaps even more significant. The chromospheric phenomena, the flares, the prominences, etc., look strangely like electrical discharges in a gas (although such a simple explanation does not hold quantitatively) and have certainly something to do with the behaviour of plasmas in magnetic fields (those of the sunspots). More puzzling still is the fact that, in the solar atmosphere, the temperature generally goes up with altitude, reaching about one million degrees in the tenuous corona. This indicates conclusively that the energy is carried upwards in a non-thermal form.

In the solar corona there is a general upwards motion of particles accelerated by complicated and only partially elucidated plasma-physics phenomena, and recently it has been shown that this motion goes on far beyond the Earth's orbit. This 'solar wind', as it is termed, is responsible for quite a number of phenomena, both in the vicinity of the Earth and beyond. Apart from this continuous 'wind', evidence has been found of more important localized plasma blobs, related to abnormal solar activity.

Both the normal solar wind and the abnormal solar blobs interact with the magnetic field of the Earth (and probably of other planets) where new phenomena occur. It has been a major discovery of recent years that important plasma rings, the so-called Van Allen belts, are trapped in the magnetic field of the Earth, and probably other planets (however, Jupiter is the only one for which positive evidence has been conclusively demonstrated). Although part of those belts can confidently be traced to a component of cosmic radiation (the so-called albedo neutrons, which are reflected by the Earth's atmosphere and decompose into a proton and an electron at some distance), it is certain that there is interaction between the radiation belts and the solar wind. When plasma blobs succeed in travelling down the magnetic lines of force of the Earth, they interact with the upper atmosphere, creating phenomena akin to electrical discharges, termed polar aurorae.

It is well known that the upper layers of the terrestrial atmosphere are ionized (the ionosphere); most of this ionization can be traced to the different radiations emitted by the sun. Minor components can be traced to cosmic radiation, meteorites, and particles escaping downwards from the Van Allen belts. The importance of the ionosphere for radio communications is well known: most of the ionosphere acts as a mirror, but the lower layers are more or less absorbing. In case of abnormal solar activity, those absorbing layers can be temporarily reinforced, causing fading out of radio transmissions as the result of a stronger absorbent being interposed between us and the mirror.

It has been realized that the tails of comets are plasma-physics phenomena. Their orientation away from the sun is attributed to the effect of the solar wind, though for some tails the older explanation involving the pressure of light on tenuous solid particles still holds good.

As we go deeper into space we continue to find many manifestations of the plasma state and of cosmic magnetic fields; but unfortunately, for the most part satisfactory interpretations are missing so far. Magnetic stars have been observed, with magnetic fields of the same order as that of the sunspots; and what is more, such magnetic stars undergo very considerable modifications in a very short time.

All the gaseous nebulae are, at least partially, in the plasma state. Their luminosity is due to photo-excitation from neighbouring hot stars emitting significant amounts of ultraviolet light. But among the gaseous nebulae, a special class has received much attention from plasma physicists; the so-called supernovae remnants. The Crab Nebula is the most conspicuous example of a supernova remnant and has been studied very thoroughly. Evidence has been found of phenomena typical of very hot plasmas in magnetic fields: synchrotron radiation, for instance.

Since the advent of artificial satellites has enabled us to explore the radiation coming from outer space throughout the range of wavelengths, more and more evidence has been found of 'mysterious' sources of energy in the galaxy and in other galaxies, too. Although unquestionable explanations are generally so far still lacking, there is little doubt that some unknown or little known aspect of plasma physics is responsible for them.

Part Five **Relativity**

Einstein's theories of relativity are based on simple concepts, but if one follows them through in detail they can involve some very complex mathematics. However it is possible to explain several very important parts of the theory without too much detailed mathematics as I hope is shown by the papers I have selected. The special theory of relativity deals with the laws of physics as they apply to observers in inertial frames of reference. By an inertial frame of reference is meant one which is not accelerated or subject to gravitational forces. If two inertial frames are in relative motion the special theory shows that measurements of distances and times made by an observer in one frame are not the same as those made by an observer in the other frame. The theory of general relativity deals with non-inertial frames of reference.

Probably the best-known conclusion of the theories of relativity is Einstein's famous equation showing the equivalence of energy and mass, $E = mc^2$. Its use to explain the large amounts of energy released in nuclear fission is well known. The reader may have come across other snippets of relativity theory also: the increase of mass with velocity, or the slow running of clocks, or more likely, the twin paradox. But I will not dwell on any of these at this stage as they are all covered in the papers which follow.

The subject is best approached for the first time in a way which strips it of all its complex mathematics and allows the student to see the scope of the theory and the way in which it applies in certain cases. This is the method adopted by M. M. Woolfson in his paper. However, even in this simplified form the work needs very careful reading and re-reading, but there is nothing in it which should be beyond the reader. Some of the conclusions may seem difficult to accept in the first instance but it must be borne in mind that conceptions of what is right or makes sense are based on everyday observations, and that all such observations involve velocities which are small compared with the velocity of light. It is only at velocities approaching that of light that the predictions of the theory of relativity do not agree with many preconceived ideas of space and time.

The two papers by O. R. Frisch deal with one aspect of relativity theory: the behaviour of time. Part I is confined to special relativity and Part II extends to general relativity.

13 M. M. Woolfson

Non-Specialist Relativity

M. M. Woolfson, 'Non-specialist relativity', *School Science Review*, vol. 49, 1968, no. 169, pp. 704–24.

1 Introduction

Relativity theory can often seem to be conceptually difficult, to require sophisticated mathematics and to complicate our description of physical laws. This is not really so; without the theory (or another giving similar results) we should find numerous baffling paradoxes to undermine all those other theories which explain the physical universe so well. Again, many of the important ideas of relativity theory can be derived with easily understood mathematics.

Perhaps the most astonishing conclusion from relativity theory is the so-called 'twin paradox', illustrated in Figure 1. If one of a pair of twins made a high-speed round trip through space he would find on his return that he was younger than his earthbound brother!

Our objective here is to give simple derivations of some of the more important results of special relativity and also to explain the 'twin paradox'.

2 Viewpoints

An incident has occurred in inter-stellar space involving two spaceships, A and B, with another spaceship, C, nearby. The captains' reports follow:

Captain A

My spaceship was stationary with its ion-drive motors switched off. We observed spaceship B travelling at 200 km s^{-1} on an attacking course whose nearest approach was 100 km from us. When he was 224 km away I fired a warning missile at 100 km s^{-1} perpendicular to his path which just missed him as he

the departure 1968

the return 2020

Figure 1

passed the point of nearest approach. He then accelerated away from the area. My report is illustrated in Figure 2.

Captain B

My spaceship was stationary with its ion-drive motors switched off. We observed spaceship A travelling at 200 km s^{-1} on an attacking course whose nearest approach was 100 km from us. When he was 224 km away he fired a missile at 224 km s^{-1} at $26\frac{1}{2}°$ to his line of motion which just missed us as he passed the point of nearest approach. I then accelerated away from the area. My report is illustrated in Figure 3.

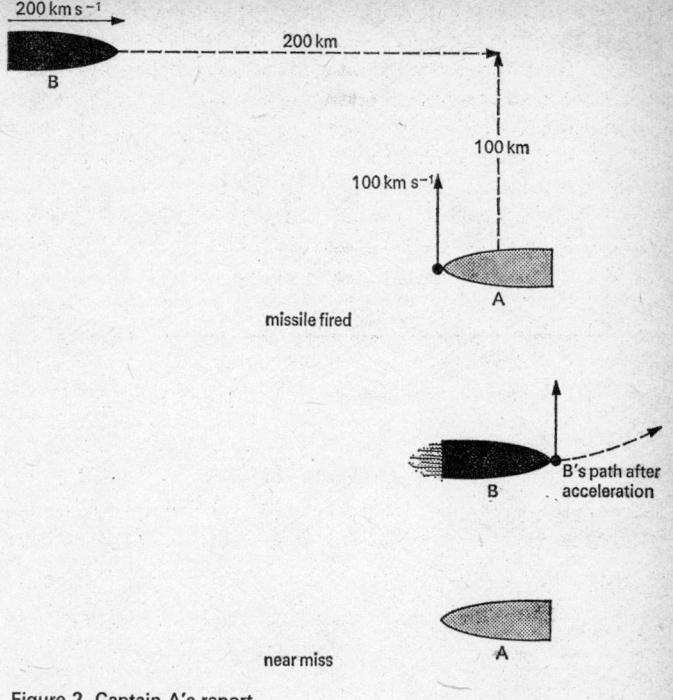

[Figure: Captain A's report diagram showing B at 200 km s⁻¹ moving right, 200 km away; A below with missile fired at 100 km s⁻¹ upward from 100 km separation; then near miss with B's path after acceleration]

Figure 2 Captain A's report

Captain C

My spaceship was stationary with its ion-drive motors switched off. We observed spaceships A and B both travelling at 100 km s^{-1} on parallel paths which were 100 km apart. When the spaceships were 224 km apart A fired a missile at 141 km s^{-1} at 45° to his line of motion. This just missed B when the two spaceships were closest together. Then B accelerated away from the scene of the incident. My report is illustrated in Figure 4.

What do we make of this apparently conflicting evidence? The observers are reporting honestly although they ascribe false motives to each other. Each spacecraft is drifting along and is not subjected to any forces, and each captain considers himself

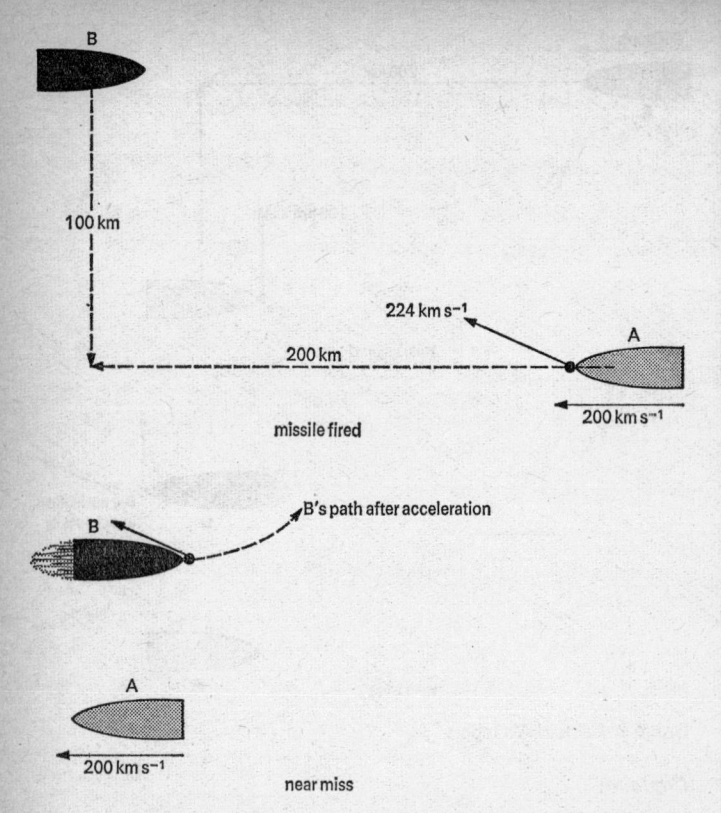

Figure 3 Captain B's report

to be a suitable, unbiased frame of reference for describing outside events. A frame of reference which *experiences* no forces is referred to as an inertial frame of reference and each of the spacecraft is such a frame of reference (except B when it accelerated for then it *experienced* a force).

We thus note that inertial frames are not all *equivalent* for describing physical phenomena but they are *equivalently valid* and the same scientific laws are seen to apply from each of them.

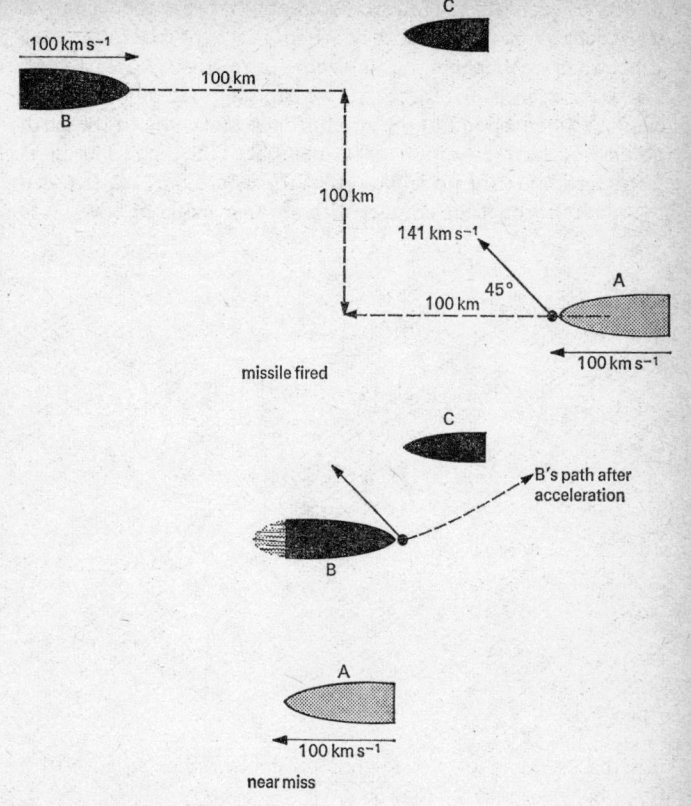

Figure 4 Captain C's report

3 Frames of reference

An inertial frame of reference is defined as one which experiences no forces because any observer (or recording device) associated with the frame is in a state of free fall. A common misconception about inertial frames of reference is that any two of them must be in a state of constant velocity with respect to each other. In Figure 5 there are two frames of reference, S_1 and S_2, represented by cartesian-coordinate axes. The gravitational fields due to m_1, m_2 and m_3 cause S_1 and S_2 to be in a state of acceleration with respect to each other. However, we note that *both* the frames of reference are inertial since they are in a state of free fall.

Is a man on earth an inertial frame of reference? Clearly not, for he experiences the force that we call the force of gravity. He cannot freely respond to the gravitational attraction of the earth because of the intervention of the material of the earth. This illustrates an important principle – that forces are only experienced by a body through its contact with another material body. It is

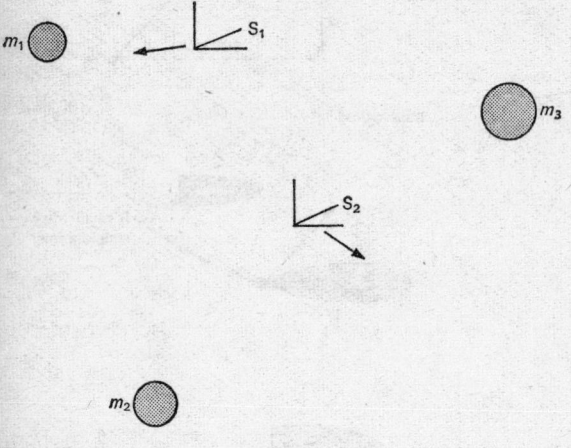

Figure 5

not the existence of a field that produces the experience of a force but resistance to that field.

Nevertheless, two men standing together on earth are equivalent *non*-inertial frames of reference. They are both perfectly good frames for observing the universe as long as they take account of the force of gravity which is, after all, just another aspect of the laws of physics.

However, the simplest frame of reference is the inertial frame and we shall consider how observers in different inertial frames moving at a constant relative velocity see the same events. We shall always assume that fields of force are absent, or alternatively are uniform, so that bodies moving at a constant relative velocity remain in that condition.

4 The speed of light

In 1867 Maxwell evolved the dynamical theory of the electromagnetic field which predicted the existence of electromagnetic waves with speed

$$c = \frac{\text{electromagnetic unit of charge}}{\text{electrostatic unit of charge}} = 3 \times 10^{10} \text{ cm s}^{-1}.$$

This value equalled the known speed of light which was thereby identified as an electromagnetic radiation.

However, all previously known types of wave (e.g. water or

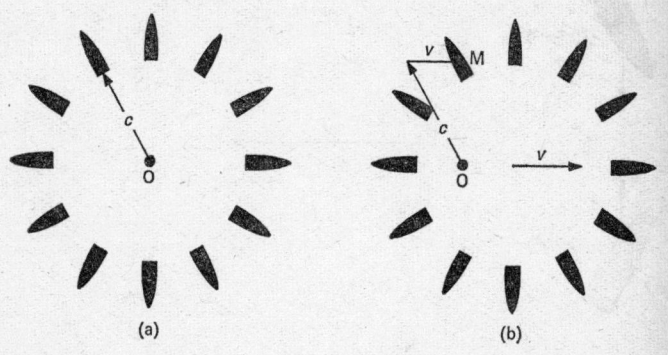

Figure 6

sound waves) travelled in a material medium, whereas electromagnetic waves were known to travel through empty space from the most distant stars. Therefore a medium was postulated, the ether, which permeated all space and in which electromagnetic waves could travel. To explain the high velocity of light this medium was assumed to have very low density and very high elasticity. If Maxwell's theory gave the speed of light relative to the ether then the speed measured by an observer would depend on his velocity with respect to the ether.

The principle of a method which was used to try to detect an observer's motion through the ether is illustrated in Figure 6. A man, O, on a small island in the middle of a large lake of still water, launches a number of identical toy boats at a speed c in various directions. After unit time the boats have reached the positions shown in Figure 6(a). The experiment is repeated when the water is flowing with a velocity v ($v < c$) and now, after unit time the boats have reached the positions shown in Figure 6(b). We should note that for a boat to reach a position M, such that OM is perpendicular to v, the direction of launch is such that the actual speed along OM is $\sqrt{(c^2-v^2)}$.

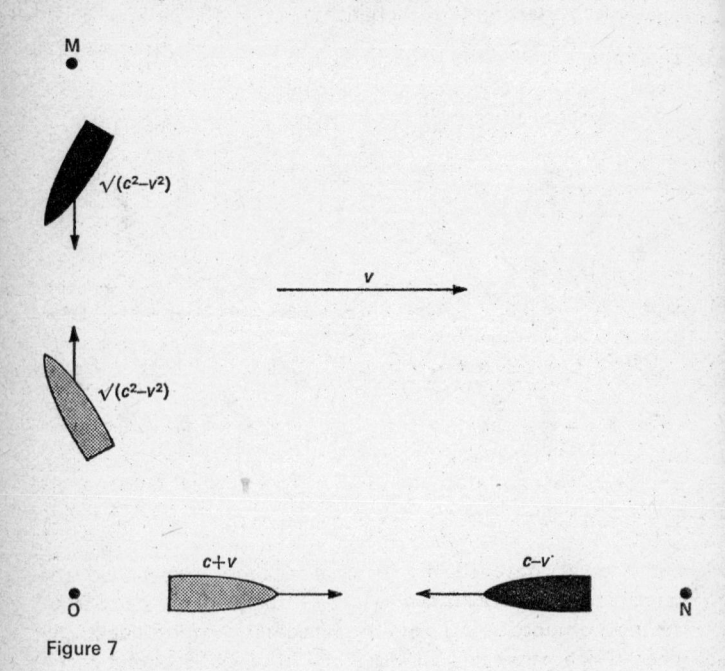

Figure 7

Now suppose there are three men, O, N and M, such that $ON = OM = l$ and ON is perpendicular to OM (Figure 7). The man at O simultaneously launches boats to M and N who, when they receive them, immediately send them back again. Which boat returns first?

The time for a boat to go from O to M and back again is the time taken to travel a distance $2l$ at a speed $\sqrt{(c^2-v^2)}$ or

$$t_{OMO} = \frac{2l}{\sqrt{(c^2-v^2)}} = \frac{2l}{c} \frac{1}{\sqrt{(1-v^2/c^2)}}. \qquad 1$$

The boat travelling from O to N and then from N to O does so at speeds $c+v$ and $c-v$ respectively, so that

$$t_{ONO} = \frac{l}{c+v} + \frac{l}{c-v} = \frac{2l}{c}\frac{1}{1-v^2/c^2}. \qquad 2$$

Since $1-v^2/c^2 < 1$, then $\sqrt{(1-v^2/c^2)} > 1-v^2/c^2$

and

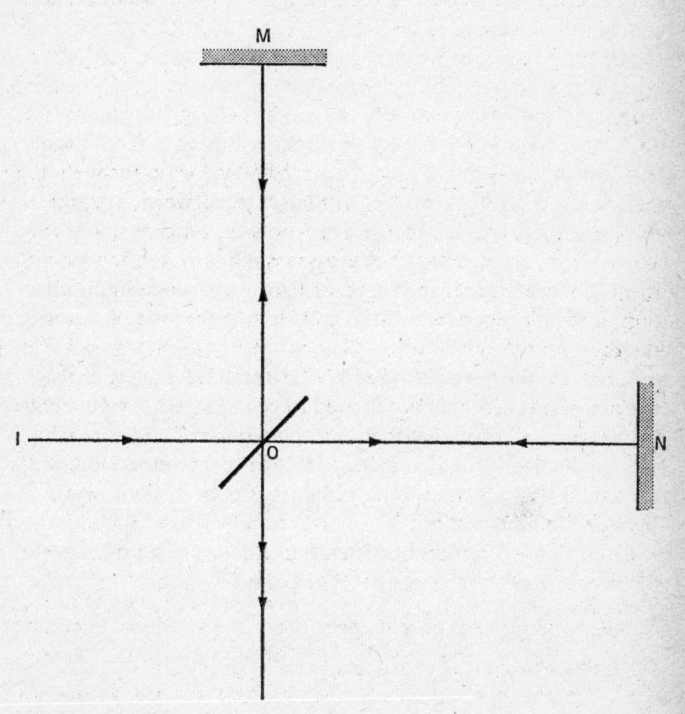

$$t_{OMO} < t_{ONO}. \qquad 3$$

Figure 8

In 1887 Michelson and Morley performed an optical analogue of this experiment with the interferometer shown in Figure 8. An incident beam of light IO was split into two by the half-silvered mirror at O, reflected by M and N and then recombined by O.

The recombined beams interfered with each other in a way which depended on the time difference for the journeys OMO and ONO. Some part of this difference may have been due to slight differences in path length but the remainder should have been due to the motion of the apparatus through the ether. The apparatus was then rotated through 90°, thus interchanging the directions of ON and OM and presumably also reversing that part of the time difference due to travel through the ether. This rotation should have produced a change in the interference pattern but no such change was observed. The negative result might have been due to the earth being, by chance, momentarily at rest relative to the ether so the experiment was repeated six months later when the velocity of the earth relative to the sun had changed by 60 km s^{-1}. Even then no positive result was obtained although, theoretically, the instrument should have recorded a velocity through the ether as low as 10 km s^{-1}. Although there were attempts to reconcile the existence of the ether and the Michelson–Morley experiment, the idea of the ether was eventually abandoned and it was concluded that the speed of light is the same in all directions and is independent of the motion of either the observer or the source.

What would have followed if the result of the Michelson–Morley experiment had not proved to be negative? It would have led to the conclusion that there was one frame of reference which was distinctive – the one at rest relative to the ether. For with respect to this one, and to no other, the speed of light would be the same in all directions.

From this background Einstein put forward the postulates of the special theory of relativity. These are:

(i) All inertial frames of reference are equally valid for describing events in the universe and the laws of physics are the same as seen from any one of them.

(ii) There is no experiment which can distinguish one inertial frame from another and, in particular, the speed of light is the same in all inertial frames.

There is a certain philosophical wholesomeness in these postulates. If, for example, someone suggested that the account given by Captain A in section 2 was more valid than the others then we should demand to know why; anything which suggested that one inertial frame of reference was better than any other would tend to make us feel a little disturbed.

So we are philosophically content with the postulates of the special theory of relativity – but where do they lead us scientifically?

5 Length and time!

Two men, A and B, possessing identical clocks set up a row of markers at an agreed spacing. While A stays by a light source S, B rides in a trolley at a constant speed v as shown in Figure 9.

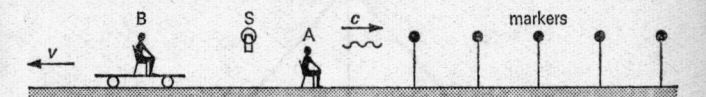

Figure 9

The light source then emits a brief light pulse which illuminates the markers in turn as it passes by. When A divides the distance between markers by the time interval for light to travel between them, he obtains the expected speed of light.

But what does B observe? He is a perfectly good frame of reference, equivalent to A, and relative to him the light pulse is moving away with a speed c (basic postulate of special theory) while the markers are moving away with a speed v. Thus the relative speed between the light and the markers as seen by B is $c-v$ and consequently he should record a time interval between the illumination of successive markers longer than that recorded by A! Clearly this conflicts with our instinctive ideas about length and/or time and we shall see that to restore sense to the situation the 'sacred absolutes' of length and time must be sacrificed.

6 A moving clock runs slowly

In Figure 10 there is illustrated a hypothetical experiment which shows that observers in relative motion have different time scales. By this we do not simply mean that one measures time in seconds while the other uses some other unit – for example microdays. Rather we mean that the two observers, using equivalent clocks, will record different times for the interval between two events – for example twenty-five seconds and thirty seconds – and that between the two events they will *experience* different intervals of time.

Two observers, provided with identical clocks, are situated at distance d from a mirror. A short pulse of light is emitted along OM and simultaneously O′ sets out parallel to the mirror at a speed v which just brings him to the point P with the light pulse. We can agree that O″'s clock can only be seen when the

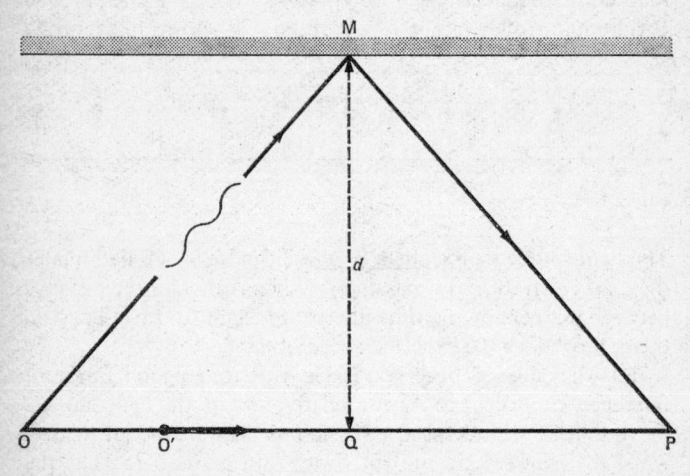

Figure 10

pulse of light falls on it so that O and O′ must agree about the time interval indicated by O″'s clock while the light makes the journey OMP.

The time interval observed by O on his own clock is t_{oo} and, since $OQ = \frac{1}{2}vt_{oo}$, he records the total path of the light pulse as

$$l_0 = 2\sqrt{[(\tfrac{1}{2}vt_{oo})^2 + d^2]}. \qquad 4$$

But the speed of light seen by O must always be c so that

$$l_0 = ct_{oo} = 2\sqrt{[(\tfrac{1}{2}vt_{oo})^2 + d^2]} \qquad 5$$

or $t_{oo} = \dfrac{2d}{c\sqrt{(1-v^2/c^2)}}.$ \qquad 6

However as seen by O' the pulse of light travels perpendicular to the mirror (the component of the velocity of the pulse parallel to the mirror is always v). Thus the time interval recorded by O' on his own clock, $t_{O'O'}$, which is the same as the time interval recorded by O on O''s clock, $t_{OO'}$, is given by

$$t_{O'O'} = t_{OO'} = \frac{2d}{c}. \qquad 7$$

From **6** and **7** we find

$$t_{OO'} = t_{OO} \sqrt{(1 - v^2/c^2)}, \qquad 8$$

which, since $\sqrt{(1-v^2/c^2)} < 1$, means that O sees a lesser time interval recorded on O''s clock than he sees on his own between the same two events. This may be summed up in the phrase 'a moving clock runs slowly'. Although our analysis tells us nothing about what O' sees on O's clock it is clear from symmetry arguments that to O' it is O's clock which is running slowly.

Let us see what this means in real terms. If O and O' were both smoking cigarettes then each would think that his own cigarette was lasting the normal time (say ten minutes) while the other man's lasted fifteen minutes. There would be no change in the experience of time by either observer. They would breath at the normal rate, age at the normal rate and the rates of chemical reactions and radioactive decay for materials at rest relative to them would be normal. It is only the other man who would see the slowing down of all these processes.

An interesting example of a moving clock which runs slowly is seen in the arrival of π-mesons at the earth's surface. These are produced by the action of cosmic rays on atoms at a height of about 20 km (2×10^6 cm) and they travel towards the earth at speeds approaching c. In the laboratory slowly moving π-mesons decay with a half-life of $2 \cdot 55 \times 10^{-8}$ s, so that even at the speed of light one half of the π-mesons should decay for every 800 cm of travel. By the time they reach the surface of the earth they should have travelled $2 \times 10^6/800$ half-life times so that the fraction of those that are produced that reach the earth should be

$(\frac{1}{2})^{2 \cdot 5 \times 10^3} \simeq 10^{-750}$.

Estimates of the number of atoms in the universe are in the range 10^{79} to 10^{80} but 10^{750} π-mesons should be produced for even one to reach the earth! The answer to this problem is that we can regard the π-mesons as a moving clock, their rate of decay being the means of measuring time. To an observer on earth the meson clock is running slowly by a factor $\sqrt{(1-v^2/c^2)}$ and so large numbers of them are able to reach the earth's surface.

7 The twin paradox again

Let us think again about the twins. The earthbound twin could see his brother making both the outward and inward journey at a speed v and so would observe his brother's ageing processes slowed down by a factor $\sqrt{(1-v^2/c^2)}$. This would explain the age discrepancy as far as he was concerned. But, we could argue, if all motion is relative why does the space traveller not observe that it is the earthbound brother who is younger? This is the paradox and it is one we shall resolve.

8 Moving bodies are shorter

In Figure 11 is shown a rod SM having at one end a light source and at the other a mirror. A short pulse of light travels from S to

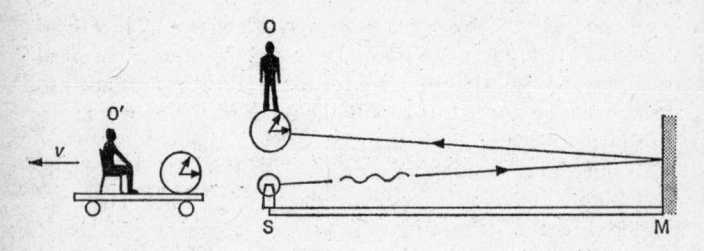

Figure 11

M and back again. An observer O at rest relative to the rod measures the length of the rod as l and records the time of passage of the light pulse on his own clock as

$$t_{OO} = \frac{2l}{c}.$$ 9

An observer O′ moving relative to the rod, along the direction of its length, at a speed v will record the same time of passage on

O's clock (imagine O's clock is only seen when the light pulse illuminates it) but a longer time on his own clock.

$$t_{O'O} = \frac{t_{O'O}}{\sqrt{(1-v^2/c^2)}} = \frac{t_{OO}}{\sqrt{(1-v^2/c^2)}} = \frac{2l}{c\sqrt{(1-v^2/c^2)}}. \qquad 10$$

Now *to O'* the speed of the light pulse is always c so when it is going along SM its speed relative to the rod is $c-v$ whereas when it is going along MS its speed relative to the rod is $c+v$. Let the length of the rod be l' as seen by O'. Then

$$t_{O'O'} = \frac{l'}{c-v} + \frac{l'}{c+v} = \frac{2l'}{c(1-v^2/c^2)}. \qquad 11$$

From 10 and 11 we find

$$l' = l\sqrt{(1-v^2/c^2)}, \qquad 12$$

or that the length of the rod appears less to O' because of his motion along the direction of its length.

Let us return to our π-mesons for, as we left our description of the phenomenon, a point of doubt could arise. An observer travelling *with* the π-mesons would record them decaying at laboratory-measured rate, so how does this observer explain that they reach the surface of the earth? The answer is that this observer is moving relative to the atmosphere so that to him it is not a 20 km but only a $20\sqrt{(1-v^2/c^2)}$ km journey to the surface of the earth. The observer on earth and the moving observer now agree that some π-mesons will reach the earth's surface, although one says it is because distances vary with relative speed while the other says that it is because time intervals vary with relative speed.

9 Combining velocities

'Nothing can go faster than light.' This is a statement which is frequently made, and in one way is true, but which requires elaboration. In fact one can have the relative speeds of two bodies greater than c as long as the observer who judges the relative speeds is not one of the bodies. If, for example, we fire two pulses of electrons in opposite directions each with a speed of $\frac{3}{4}c$ then the relative speed, as we measure it, is $\frac{3}{2}c$. Again, if one points a laser at the moon which is at a distance of 400 000 km and rotates the laser at 1 rad s^{-1} then the patch of illumination sweeps over the surface of the moon at 4×10^{10} cm s^{-1} which is greater than the speed of light. What *is* true is that we cannot observe either material or information transmitted at greater than the speed of light.

In the situation shown in Figure 12(a) an observer O (represented by a clock) records the passage of a rod AB which, when at

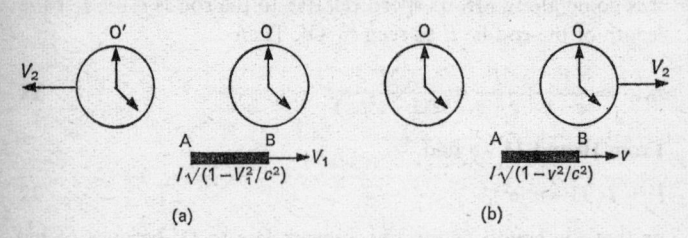

Figure 12

rest relative to an observer, has a length l. The time recorded by O on his own clock between the passage of A and B is

$$t_{OO} = \frac{l\sqrt{(1-V_1^2/c^2)}}{V_1}, \qquad 13$$

since, for O, the rod has length $l\sqrt{(1-V_1^2/c^2)}$.

The time recorded by O' on O's clock $t_{O'O}$ must equal t_{OO}, by arguments we have used before, so that the time he records on his own clock is greater and is

$$t_{O'O'} = \frac{t_{OO}}{\sqrt{(1-V_2^2/c^2)}} = \frac{l\sqrt{(1-V_1^2/c^2)}}{V_1 \sqrt{(1-V_2^2/c^2)}}. \qquad 14$$

Now let us look at the situation from O"s frame of reference as in Figure 12(b). The observer O is moving away at a speed V_2 and the rod is moving away at a speed v which, by the classical rules of combining velocities, would be $V_1 + V_2$. But we know that our relativity answer must be different, for $V_1 + V_2$ could be greater than c and O' cannot see a material body moving relative to *himself* at a speed greater than c. We can now find an expression for v.

The length of the rod to O' is $l\sqrt{(1-v^2/c^2)}$ and the relative

268 Relativity

speed of rod and O is $v - V_2$. Hence the time interval observed by O' on his own clock between the passage of A and B past O is

$$t_{\text{O'O'}} = \frac{l\sqrt{(1 - v^2/c^2)}}{v - V_2}. \qquad \textbf{15}$$

Equating the two expressions for $t_{\text{O'O'}}$ in **14** and **15** gives a quadratic equation for v with solutions

$$v = \frac{V_1 + V_2}{1 + V_1 V_2/c^2} \quad \left(\text{or } v = \frac{V_1 - V_2}{1 - V_1 V_2/c^2}\right). \qquad \textbf{16}$$

These solutions replace the classical expressions

$$v = V_1 + V_2 \text{ (or } v = V_1 - V_2\text{)},$$

the intervening sign depending on the relative directions of V_1 and V_2.

The expression **16** has the interesting property that if either of the velocities is c then the combined velocity is c. For example with $V_1 = c$

$$v = \frac{c + V_2}{1 + V_2/c} = c,$$

and this agrees with the result that observers moving relative to one another all obtain the same value for the speed of light.

The other property of interest is that no combinations of speeds less than c can give a resultant greater than c. Thus with $V_1 = V_2 = 0 \cdot 8c$

$$v = \frac{0 \cdot 8c + 0 \cdot 8c}{1 + 0 \cdot 64c^2/c^2} = 0 \cdot 976c.$$

We ought to be quite clear that there are now two rules for combining speeds to give relative speeds. The first is when an observer O combines the speed of object A, V_A, with the speed of object B, V_B, both relative to himself, to obtain the relative velocity of A and B, $V_A \pm V_B$ according to the classical formula. The other is when an observer combines the speed of an object A, V_{OA}, relative to himself and of an object B, V_{AB}, relative to A to obtain the resultant V_{OB} which must be done through equation **16**.

10 Moving masses become greater

Consider the situation shown in Figure 13(a) where two equal particles of mass m move with equal speeds V relative to observer O, collide head on, adhere together and remain at rest. The

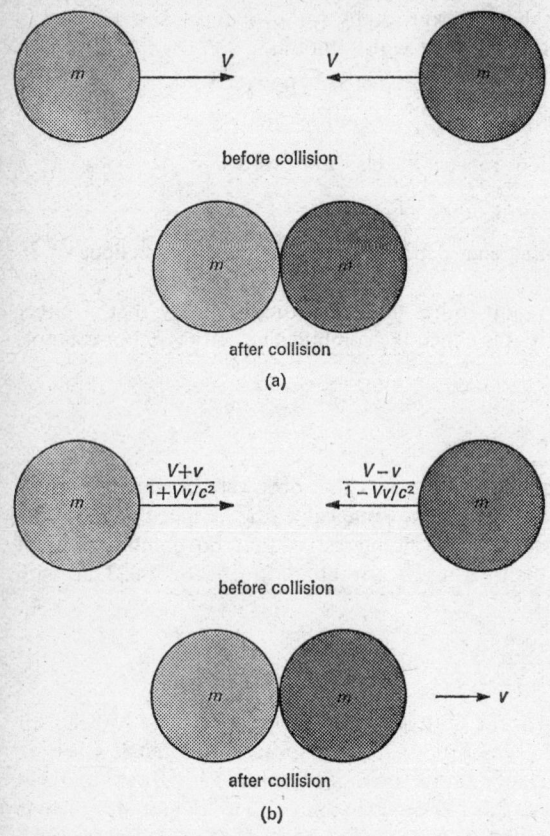

Figure 13

momentum before the collision is $mV - mV = 0$ and remains zero after the collision.

Another observer, O', moving at a speed v relative to O, sees the initial speeds as in Figure 13(b), that is

270 Relativity

$$V_1 = \frac{v+V}{1+vV/c^2} \quad \text{and} \quad V_2 = \frac{v-V}{1-vV/c^2},$$

and after the collision the speed of the combined particles is v. However,

$$m\frac{v+V}{1+vV/c^2}+m\frac{v-V}{1-vV/c^2} \neq 2mv, \qquad 17$$

and in O''s frame of reference the principle of the conservation of momentum has broken down.

This conservation law is such a cornerstone of physics that we are loath to see it lost and we define a variation of mass with speed which enables us to retain the law. In Figure 14(a) two equal spheres undergo an elastic collision each being deflected through a right angle. The masses and speeds seen by an observer O are indicated and it is clear that momentum is conserved in this collision process. In Figure 14(b) we now consider the speeds for another observer, O', moving at a speed v with respect to O; these are indicated together with approximations to those speeds based on the condition $v \ll V$. If the mass of the sphere is m when its speed is V then it is $m+(dm/dV)\delta V$ for a speed $V+\delta V$ – that is to say that the change of mass equals the rate of change of mass with speed times the change of speed. The masses for the observer O' are indicated in square brackets.

After collision each sphere has a component of velocity v along the original direction of motion so, conserving momentum along that direction we obtain

$$\left[m+\frac{dm}{dV}v\left\{1-\frac{V^2}{c^2}\right\}\right]\left[V+v\left\{1-\frac{V^2}{c^2}\right\}\right] - \left[m-\frac{dm}{dV}v\left\{1-\frac{V^2}{c^2}\right\}\right] \times$$
$$\times \left[V-v\left\{1-\frac{V^2}{c^2}\right\}\right] = 2\left(m+\frac{dm}{dV}\frac{v^2}{2V}\right)v \qquad 18$$

which, on expansion, gives

$$\frac{dm}{m} = \frac{V\,dV}{c^2(1-V^2/c^2)}. \qquad 19$$

Integrating

$$\log m = -\tfrac{1}{2}\log(1-V^2/c^2)+\text{constant},$$

$$\frac{V-v}{1-Vv/c^2} \simeq V-v\left[1-\frac{V^2}{c^2}\right]$$
$$\left[m-\frac{dm}{dV}v\left\{1-\frac{V^2}{c^2}\right\}\right]$$

$$\sqrt{(V^2+v^2)} \simeq V+\frac{v^2}{2V}$$
$$\left[m+\frac{dm}{dV}\frac{v^2}{2V}\right]$$

$$\frac{V+v}{1+Vv/c^2} \simeq V+v\left[1-\frac{V^2}{c^2}\right]$$
$$\left[m+\frac{dm}{dV}v\left\{1-\frac{V^2}{c^2}\right\}\right]$$

$$\sqrt{(V^2+v^2)} \simeq V+\frac{v^2}{2V}$$
$$\left[m+\frac{dm}{dV}\frac{v^2}{2V}\right]$$

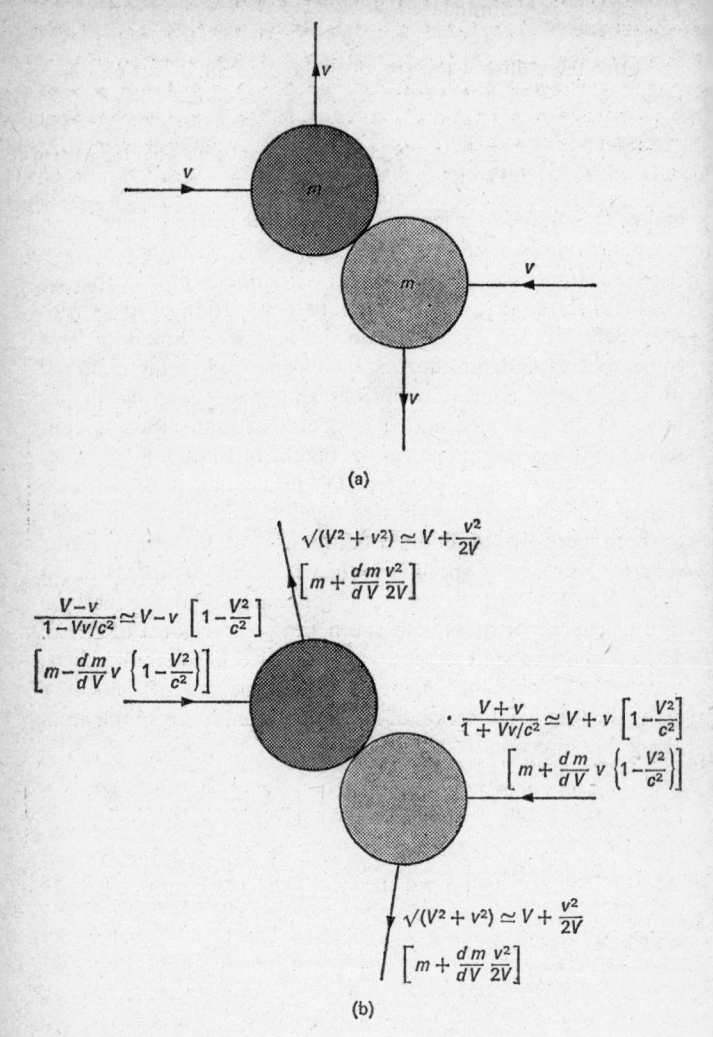

Figure 14

and if $m = m_0$ when $V = 0$ then the constant $= \log m_0$.
This gives

$$m = \frac{m_0}{\sqrt{(1 - V^2/c^2)}}, \qquad 20$$

which is the required law of variation of mass with speed. The quantity m_0 is called the rest mass of the body.

This law of variation of mass with velocity gives a consistent set of physical laws seen from all frames of reference. Thus if we deflect a beam of very fast electrons with an electric field the deduced value of e/m is consistent with a mass for the electron given by equation **20**.

11 Mass and energy

We now look at the problem of finding the total energy required to accelerate a particle from rest to a velocity V. When its velocity is v its momentum is

$$mv = \frac{m_0 v}{\sqrt{(1 - v^2/c^2)}}. \qquad 21$$

The accelerating force equals the rate of change of momentum or

$$F = \frac{d}{dt}\left[\frac{m_0 v}{\sqrt{(1 - v^2/c^2)}}\right] = \frac{m_0}{(1 - v^2/c^2)^{\frac{3}{2}}} \frac{dv}{dt}.$$

The work done in moving the particle through a distance dx is thus

$$dE = F\,dx = \left[\frac{m_0}{(1 - v^2/c^2)^{\frac{3}{2}}}\right]\frac{dv}{dt}\,dx = \left[\frac{m_0}{(1 - v^2/c^2)^{\frac{3}{2}}}\right]\frac{dx}{dt}\,dv$$

$$= \frac{m_0 v}{(1 - v^2/c^2)^{\frac{3}{2}}}\,dv.$$

Thus the total increase in energy of the particle in accelerating to a velocity V is

$$E = \int_0^V \frac{m_0 v}{(1 - v^2/c^2)^{\frac{3}{2}}}\,dv = \left[\frac{m_0 c^2}{\sqrt{(1 - v^2/c^2)}}\right]_0^V = mc^2 - m_0 c^2, \qquad 22$$

where m is the mass at a velocity V. This can also be written as

$$\Delta E = \Delta mc^2, \qquad 23$$

relating a change of energy to a change of mass. It is inferred that this relationship can be applied whenever, in some process, mass disappears and energy is produced. This inference is now well established as valid in the production of energy from nuclear reactions.

Equation 22 is occasionally interpreted as meaning that a total amount of energy $E = mc^2$ is associated with a mass m and it is in this form that the law relating mass and energy is usually stated – although it is not easy rigorously to justify this relationship.

We should notice that, when $V \ll c$, we have

$$\Delta E = m_0 c^2 \left[\frac{1}{\sqrt{(1 - V^2/c^2)}} - 1 \right] = \tfrac{1}{2} m_0 V^2, \qquad 24$$

the classical expression for kinetic energy.

It is interesting to note that the intensity of radiation from the sun falling on Earth is 1·4 kW m^{-2} which implies a total power output of approximately 4×10^{23} kW or 4×10^{26} J s^{-1}. Hence the loss of mass of the sun per second is

$$\Delta m = \frac{\Delta E}{c^2} = \frac{4 \times 10^{26}}{9 \times 10^{16}} \text{ kg s}^{-1} \simeq 4 \times 10^6 \text{ tonne s}^{-1}!$$

12 Gravitational fields and clocks

In Figure 15 are shown two non-inertial observers, O and O′, at rest relative to the sphere of mass M. A photon of frequency v is emitted from O towards O′. The photon may be considered as a particle whose energy of motion is hv, mass hv/c^2 and momentum hv/c.

The increase in potential energy of the photon in going from O to O′ is

$$\frac{hv}{c^2} \left[\frac{GM}{r_O} - \frac{GM}{r_{O'}} \right] \qquad 25$$

and hence the energy of motion of the photon must decrease by this amount. Hence the difference in frequency seen by O, dv, is given by

$$\frac{dv}{v} = -\frac{GM}{c^2} \left[\frac{1}{r_O} - \frac{1}{r_{O'}} \right] = -\frac{\Phi_{OO'}}{c^2}, \qquad 26$$

where $\Phi_{OO'}$ is the difference in gravitational potential between O and O' (work done against the field in taking a unit mass from O to O').

If O was a point on the surface of a star and O' an observer at a great distance then the light seen by O' is lowered in frequency (reddened) by its climb up the potential gradient. This effect, the gravitational red shift, can be seen in the spectra of some stars by observations of the Fraunhoffer lines. The star Sirius has a mass equal to that of the sun, 2×10^{33} g, a fact deducible because

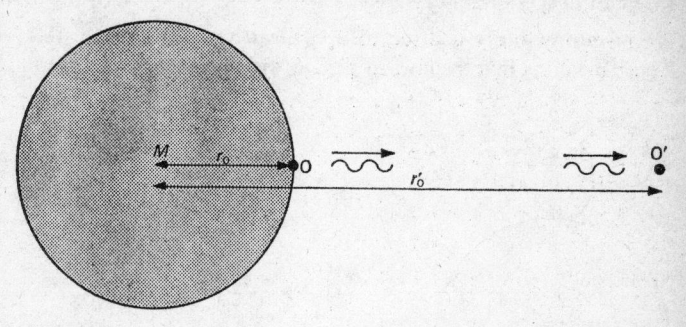

Figure 15

Sirius is one member of a binary star system. For the light from Sirius one finds $dv/v = -7 \times 10^{-4}$. From **26** with r_O as the radius of the star and $r_{O'} \gg r_O$ we find that the radius of Sirius is 2×10^8 cm which implies a density of 10^8 g cm^{-3}! Stars which exist in this superdense state are known as white dwarfs.

An atom emitting a spectral line can be considered as a clock with each cycle of the electric vector indicating an interval of time $1/v$. Thus an observer looking at an atomic clock (or indeed at any physical process with which a definite rate is associated) which is at a lower potential than himself observes the clock (or physical process) to run slowly. Once again we must appreciate that the observer does not experience any personal sense of slowness in the passage of time nor do the rates of very close processes seem different to him. This can be inferred from **26** since $\Phi_{OO} = 0$.

We should note that the gravitational red-shift phenomenon is reversible and, since $\Phi_{O'O} = -\Phi_{OO'}$, light going from O' to O will seem to be increased in frequency, corresponding to a gravitational blue shift.

In the above discussion we have non-inertial observers and we have moved into the realm of general relativity theory. Our simple Newtonian view of a gravitational field is modified by the general theory so we are not really justified in setting up equation **25**. However, for the limited application we are making the Newtonian approximation is justified.

13 Equivalent frames of reference

We have previously referred to *equivalently valid* frames of reference meaning that the laws of physics are the same as seen from

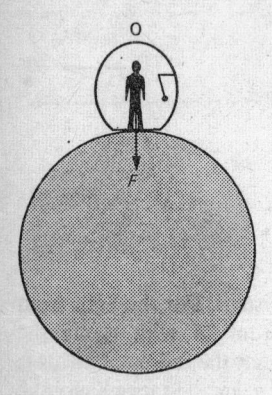

Figure 16

both frames. Now we shall consider the concept of *equivalent* frames of reference implying that the same events are seen in the same way from each of them.

In Figure 16 there are two non-inertial frames of reference represented by the observers O and O'. These observers are both capsulated in identical opaque enclosures and so cannot use visual evidence to examine their environment. Observer O is situated on earth and experiences a force $F = mg$ due to the earth's gravitational field. By experiment, for example with a pendulum, O can both detect and measure g. Observer O' is in the depths of space in a rocket ship with its motors switched on. The reaction to the rocket blast presses the floor against O' who experiences a force F (equal to his mass × acceleration with respect to an inertial frame in the same environment). If the acceleration of O' equals g he cannot distinguish his situation from that of O – a simple-pendulum experiment will give the same result as it does in a gravitational field.

Now imagine that the capsules have windows so that the observers can see some outside events but still cannot tell whether they are on a planet or in a rocket. An important proposition of general relativity, the principle of equivalence, states that the observers cannot distinguish their environments by observation

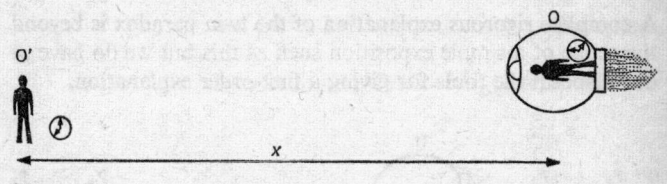

Figure 17

of outside events, which appear the same to both of them and therefore they are *equivalent* frames of reference.

Let us now consider Figure 17 where an observer O has an acceleration f towards O' with a distance x between O and O'. Then O experiences the equivalent of a gravitational field f and sees all outside bodies apparently falling freely in a *uniform* gravitational field f occupying the space around him. Thus he would judge that there was a gravitational potential difference between himself and O' given by

$$\Phi_{OO'} = fx. \qquad 27$$

From the results in section 12 we can see that O will observe a blue shift of light coming from O' or that he will see a clock at O' running faster than his own. A time interval Δt_{OO} seen by O on his own clock will be related to a time interval $t_{OO'}$ seen by O on O''s clock by

$$\frac{\Delta t_{OO'}}{\Delta t_{OO}} = \frac{v + dv}{v} = 1 + \frac{fx}{c^2}. \qquad 28$$

Notice that as O accelerates towards a clock that clock is seen to run faster by a factor linearly dependent on its distance but if

O accelerates away from a clock then the clock is seen to run more slowly. However, the acceleration of O does not have any effect on O′ who does not experience O's acceleration. While O′ will see O's clock run more slowly than his own it will only be by the factor given in equation **8** which depends on their relative velocity. While a gravitational field gives reciprocally related red and blue shifts for observers O and O′ the acceleration of O gives an equivalent effect only for O who actually experiences the acceleration.

14 The twin paradox – an explanation

A complete rigorous explanation of the twin paradox is beyond the scope of a simple exposition such as this but we do have at our disposal the tools for giving a first-order explanation.

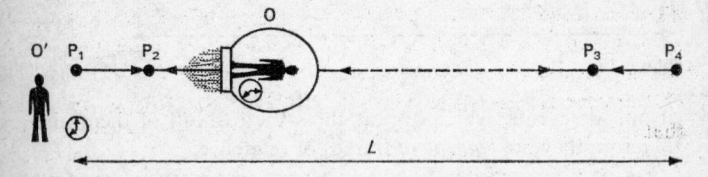

Figure 18

We consider the space journey by the twin O in five stages as in Figure 18.

1. $P_1 P_2$, rapid acceleration to speed v
2. $P_2 P_3$, steady relative speed v
3. $P_3-P_4-P_3$, acceleration to change speed to v in opposite direction
4. $P_3 P_2$, steady relative speed v
5. $P_2 P_1$, rapid acceleration to bring rocket to rest

The steady velocity part of the journey dominates the remainder so that, as far as O′ is concerned, he sees O make a journey of length $2L$ virtually all of it being at a speed v. Thus the time interval he sees on his own clock is

$$\Delta T_{O'O'} = \frac{2L}{v},$$

and on O's clock he sees an interval

$$\Delta T_{O'O} = \frac{2L}{v}\sqrt{\left[1-\frac{v^2}{c^2}\right]}.$$

If $v \ll c$ then to a first approximation we have

$$\Delta T_{O'O} - \Delta T_{O'O'} = -\frac{Lv}{c^2}, \qquad 29$$

and this is the amount by which O' sees O as *younger* than himself when O returns.

From O's point of view there are three stages when he is accelerating relative to O' and therefore will see O''s clock affected. However, stages 1 and 5 are the exact opposites of each other and the amount by which O''s clock is seen to fall behind at stage 1 will be exactly compensated by the amount by which it runs ahead at stage 5. But notice that the whole of stage 3 consists of an acceleration of O *towards* O'. During the steady velocity part of the journey (distance 2L virtually all at speed v) O will see O''s clock fall behind his own by the amount given in **29** so that

$$\Delta_v T_{OO'} - \Delta_v T_{OO} = -\frac{Lv}{c^2}, \qquad 30$$

where the subscript v represents the time intervals due to steady velocity effects.

During stage 3 let us assume that O experiences a steady acceleration f. Then the time recorded by O on his own clock during this stage is

$$\Delta_f T_{OO} = \frac{2v}{f}. \qquad 31$$

From **28** we can see that the time seen by O on O''s clock will be given by

$$\frac{\Delta_f T_{OO'}}{\Delta_f T_{OO}} = 1 + \frac{fL}{c^2}. \qquad 32$$

Combining **31** and **32** we find

$$\Delta_f T_{OO'} - \Delta_f T_{OO} = \frac{2Lv}{c^2}. \qquad 33$$

From **33** and **30** we deduce that the total amount by which O sees O''s clock ahead of his own is

$$\Delta T_{OO'} - \Delta T_{OO} = (\Delta_v T_{OO'} - \Delta_v T_{OO}) + (\Delta_f T_{OO'} - \Delta_f T_{OO}) = \frac{Lv}{c^2}.$$

34

Thus O sees O''s clock ahead of his own and he sees that O' is *older* than himself. The paradox has now been resolved since from the point of view of both of the observers O' should be older by an amount Lv/c^2. While O' will *always* see O's clock running slowly, O will see O''s clock running slowly in stages 2 and 4 but during stage 3 O''s clock is seen to run quickly at a furious rate which is enough to leave O ahead in time at the end of the journey.

15 Conclusions

While the mathematical difficulties associated with relativity theory make it unpalatable to many scientists, the idea of doing without the theory is even less palatable.

It cannot be pretended that in one short article the depths of relativity theory have been explored or even that rigour has been preserved in the various derivations. However, many of the important results have been obtained without recourse to the usual mathematical techniques, the K-calculus or the Lorentz transformation. Even where non-rigorous treatments have been used it is hoped that they show that the conclusions of relativity theory are, at least, not unreasonable.

14 O. R. Frisch

Time and Relativity I

O. R. Frisch, 'Time and relativity: Part I', *Contemporary Physics*, vol. 3, 1961, no. 1, pp. 16–27.

The questions I shall discuss are not new, nor are the answers: they can mostly be found in Einstein's early papers. But the coming of artificial Earth satellites has offered new possibilities for testing some of Einstein's predictions, and some tests have actually been carried out with the help of the recoilless gamma radiation. Furthermore Dingle has cast doubt on some consequences of Einstein's theories, and the ensuing controversy has left many people thoroughly bewildered. Hence a presentation of the behaviour of time according to the theory of relativity might be useful.

Time is of course measured with clocks, and there is a variety of them. A grandfather clock would clearly be no use in a spaceship where gravity may be absent (or greatly increased as during take-off), and even a wristwatch may be slightly affected by acceleration. But that is no essential difficulty: an atomic clock – which uses the frequency associated with some atomic transition – would be far less sensitive to acceleration, and nuclear frequencies are less sensitive still.

To appreciate what the relativity theory has to say about time, we recall what Newton said in his *Principia*: 'Absolute, true, and mathematical time, of itself and from its own nature, flows equably without relation to anything external . . . .' That assumption has been criticized because it cannot be tested: Newtonian time is ticking away all through space, as it were, but its ticks cannot be heard. However, we must remember that Newtonian physics has no speed limit; so the signal from a master clock can be sent to anywhere in space with as little delay as you please so that you can tell what time it is at any place without ambiguity.

Relativity – in a limited sense – is part of Newtonian physics.

A frame of reference relative to which any free mass point moves in a straight line at constant speed is called an inertial frame, and it was known at Newton's time that any frame that moves at constant speed in a straight line relative to an inertial frame is again inertial. That fact, that the laws of mechanics are equally valid in any two inertial frames, is sometimes called the Galilean principle of relativity. It does not embrace the propagation of light (or other electromagnetic phenomena). But from about 1887 evidence accumulated that electromagnetism ought to be included in the principle of relativity; in particular, light appeared to have the same speed c relative to all frames of reference. Various attempts were made to modify mechanics accordingly; but the first simple and consistent presentation was given by Einstein in 1905.

In this article I shall not try to present all of relativity theory, but only those features which relate to the question: How do moving clocks behave? The most striking features of that behaviour are these:

1. Two events, simultaneous by the clocks in one inertial frame, will in general not be simultaneous by the clocks in another inertial frame.

2. Of two identical clocks in relative motion, each will be observed to go slow by an observer moving with the other.

3. Of two identical clocks in a gravitational field, the one at higher gravitational potential will go faster.

These features I shall derive as directly as possible from these two basic assumptions: (1) any two inertial frames are equivalent, and (2) light always travels at the same speed. In fact it is best to combine these two into one sentence which expresses the special relativity principle: *Any two inertial frames are equivalent, also with regard to the propagation of light*.

From this one assumption we can derive the statements (1) and (2) about clocks. To derive (3) we also need the principle of the universal proportionality of weight and mass (also called the equivalence principle); this will be discussed in a second article.

The decisive difference from Newtonian physics is that we now have a speed limit: no signal can travel faster than light. So to synchronize two clocks we must make do with messengers that have finite speed. In principle, all the subsequent arguments could be carried through if we employed boys on bicycles to set the clocks, but we would have to know how their speed transforms from one frame of reference to another. So it is better to use light as a messenger; its constant speed (in vacuo) relative to all frames of reference makes all arguments much simpler.

Let us start with a simple problem: how would Jack and Mac, several thousand million miles apart, but both at rest in the same inertial frame S, synchronize their clocks? Mac might start by adjusting his clock so that it always reads the same time that he can see – through his telescope – on Jack's clock. But soon he gets a message that his clock, as seen by Jack, is several hours behind; so he splits the difference by advancing his clock by half that amount. Now both clocks behave in the same way: each, as seen from the other clock, appears late by the same amount (the time the light takes from one to the other). Hence they can now be considered synchronized according to the relativity principle.

An alternative way of synchronizing those clocks would have been to set them both by a master clock, placed half-way between Jack and Mac. The symmetry of that procedure guarantees in a transparent way the equivalence of the two clocks after setting, and the result is of course the same.

Let us now place a coordinate system in S so that Jack is in its origin, and Mac on the positive x-axis. Next we introduce two more clocks, belonging to Ed and Fred who travel along the x-axis at the speed v, with Fred in front. Since everything happens on the x-axis we can ignore the y- and z-axes and draw the usual space–time graph, Figure 1. A point on that graph means an event, and we shall look at the two events O (Ed passes Jack) and P (Fred passes Mac). By chance O and P occur at the same time, according to the clocks in S, i.e. Jack's and Mac's. But not so according to Ed and Fred (frame S'). They will have synchronized their clocks by the same procedure, say, with a master clock that moves along, half-way between them, and a glance at Figure 1 shows that by their clocks the event synchronous with O is not P but Q, which is later. Both pairs of clocks have been synchronized by the relativity principle. If one pair, say Jack and

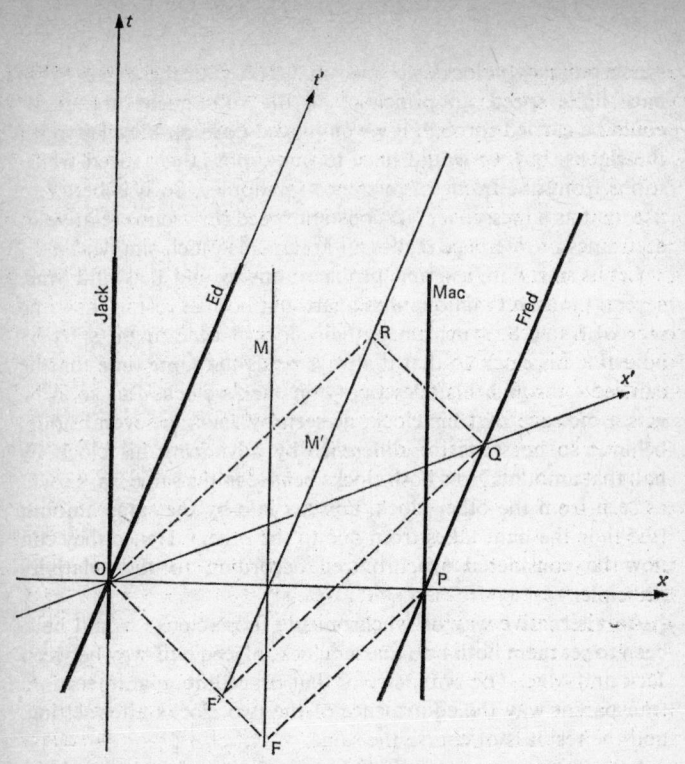

Figure 1 Relativity of simultaneity. In this space–time diagram, each point represents an event. Events at the same place in the inertial system S (the 'rest frame') lie on a line parallel to the t-axis (chosen vertical); simultaneous events lie on a line parallel to the x-axis (chosen horizontal). The travel of a light signal is shown by a dotted line.
Jack and Mac are represented by bold vertical lines, since they are both at rest in S. M is a master clock half-way between them. A light flash emitted at F will reach them at O and P; for reasons of symmetry, O and P must be simultaneous in S, confirming our choice of the horizontal direction for the x-axis.
But in S', the moving inertial system in which Ed and Fred are at rest, O and P are not simultaneous. A flash emitted at F' from the master clock M' (at rest in S' and half-way between Ed and Fred) will reach Ed and Fred at O and Q respectively; so O and Q are simultaneous in S' and Q is later than P. Fred's clock will read less as he passes Mac than Ed's clock reads as he passes Jack although Jack and Mac record those two events as simultaneous: 'The clock in front is behind in time.'
If we choose the origin of S' (i.e. the point $x' = 0$, $t' = 0$) also at O, then the x'-axis must connect O with Q, since O and Q are simultaneous in S'. It is convenient to choose length and time units so that the speed of light $c = 1$; then the 'light lines' are at 45° to the vertical and with the help of the two auxiliary lines OR and QR one sees easily that $xOx' = tOt'$

Mac, is considered at rest, then of the two moving clocks the one in front (in space) is behind (in time). Of course the relativity principle allows us to consider S' at rest and S in motion, in the opposite direction: then Jack is ahead of Mac in space and indeed again behind in time.

It is surprisingly easy to get wrong results by overlooking that 'relativity of synchronism'. For instance take the following arguments: the light energy from the flare-up of a nova is contained in an expanding spherical shell, with the nova at the centre; a body, flung off at great speed by the explosion, would not be at the centre of that shell; 'hence relative to that body the light has not been spreading at the same speed in all directions'. That conclusion contradicts the relativity principle and must be wrong. The fallacy is that the shell we mentioned contains the light energy at a given *time relative to the nova*; if we ask where the light energy is at a given *time relative to the moving body* we shall again find a spherical shell, now with that body at the centre. The motion of the nova when it exploded is quite irrelevant; relative to any inertial frame the shell is centred about the point where, *in that frame*, the nova was when it exploded.

By the way, the expanding ring around the Nova Persei of 1901 is not that shell. The shell as such can never be seen: the light from its different parts would reach us at widely different times. What we see today are those scattering particles by which the light can reach us with sixty years' delay: they form an ellipsoidal shell with us at one focus and the place (in our inertial frame) of the nova explosion at the other. That light reaches us from all directions; but it looks weak near the centre because back-scattering is weak, and falls off again with increasing distance from the explosion, giving the observed pattern of a diffuse ring.

Now to our second statement, concerning the behaviour of clocks in relative motion. Let us see how Jack and Ed would synchronize their clocks. Setting them is easy: as they pass each other they will set their clocks to the same time, say zero. But after that they will move apart and light signals will take more and more time from one to the other. Let us assume that Ed at first adjusts the rate of his clock by the receding face of Jack's clock, ignoring the growing delay. His clock then reads unit time when he sees unit time on Jack's clock; but Jack reports that he had to wait the longer time $(c+v)/(c-v)$ – see Figure 2 – before he saw unit time on Ed's clock. Once again one has to split the difference, or rather the ratio: Ed speeds up his clock by the factor $f = \sqrt{\{(c+v)/(c-v)\}}$ and resets it so that it extrapolates back to zero as before. Now the clocks are again equivalent: each clock keeper sees the receding clock go slower by the factor $1/f$.

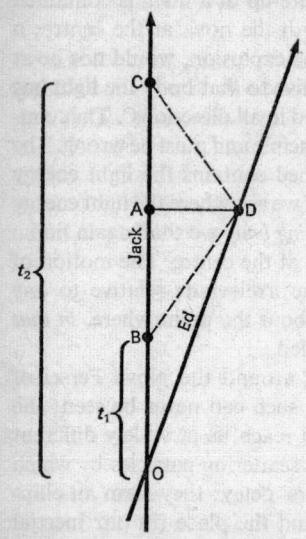

Figure 2 Relativistic Doppler effect. Both Jack and Ed have set their clocks to zero when they passed each other (event O). Some time later (event D) Ed observes that Jack's clock reads t_1 (event B) and naively he sets his clock to t_1 as well; but soon Jack reports that he had to wait until the time t_2 before (event C) he saw t_1 on Ed's clock. If the units are chosen so that $c = 1$, then

AB = AC = AD = (v/c) . AO.

It follows that

t_2/t_1 = OC / OB = $(c+v)(c-v)$.

So Ed should have adjusted his clock to a speed

$f = \sqrt{\{(c+v)/(c-v)\}}$

times higher than he did; when that is done, the two clocks are equivalent in that each appears slow by the factor $1/f$ if viewed from the other

Again, the synchronization could have been done more simply by the use of a master clock half-way in between. In order to guarantee the symmetry of the set-up we choose our rest system so that the master clock is at rest at its origin, and we assume Tim

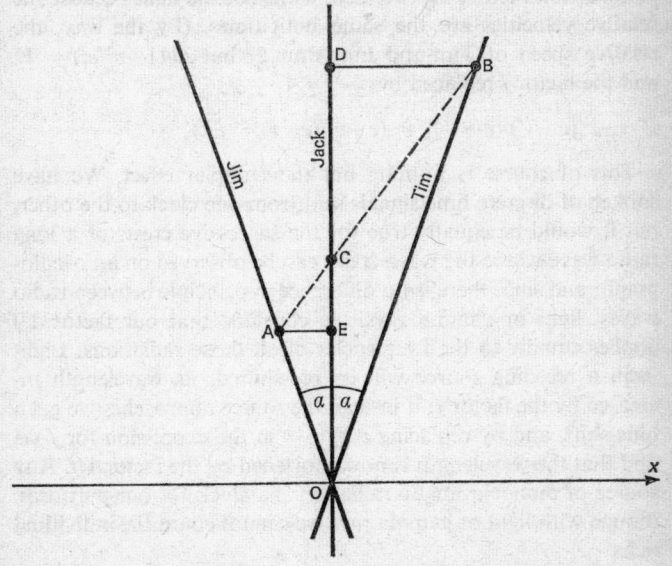

Figure 3 Another derivation of the Doppler factor f. Tim and Jim are travelling away from Jack, in opposite directions, at speed $v = c \tan a$. In Newtonian physics Jim's clock would appear slow to Jack by the factor $EO/CO = 1/(1+v/c)$, whereas Jack's clock appears slow to Tim by the factor $CO/DO = 1-v/c$; the Doppler factor here depends on whether the source or the receiver of the signals is at absolute rest. In relativity theory the two factors must be alike, but their product must be $(c-v)/(c+v)$ as before; hence the factor must be

$$\sqrt{\{(c-v)/(c+v)\}} = 1/f,$$

as derived in Figure 2

and Jim with their two clocks move away from it at speeds $+v$ and $-v$ respectively. Once we have done that we can forget the master clock: clearly the two clocks will be synchronized if their 'ticks' are represented by equal lengths in Figure 3. When Tim looks back at Jim's clock it will again read less than his own, and the ratio of the two readings is found (see Figure 3) to be $(c+v)/(c-v) = f^2$. That agrees with the result obtained in the last paragraph: Tim sees the master clock go slow by the factor $1/f$ and the keeper of the master clock sees Jim's clock go slow by that same factor; the two factors must be the same because the relative velocities are the same both times. (By the way, the relative speed of Tim and Jim is not $2v$ but $2v/(1+v^2/c^2) = V$, and the factor f replaced by

$$F = \sqrt{\{(c+V)/(c-V)\}} = (c+v)/(c-v) = f^2.)$$

This of course is nothing but the Doppler effect. We have spoken of discrete time signals sent from one clock to the other, but it would be equally true for the successive crests of a long radio wave where the wave crests can be observed on an oscillograph; and since there is no difference in principle between radio waves, light or gamma rays, we conclude that our factor $1/f$ applies equally to the frequencies of all those radiations. Light from a receding source will be red-shifted, its wavelength increased by the factor f; if instead the source approaches we get a blue-shift, and by replacing v by $-v$ in the expression for f we find that the wavelength is now shortened by the factor $1/f$. Any source of monochromatic radiation is a clock for our purposes, though with light or gamma rays we cannot count the individual ticks.

Of course the Doppler effect was known in classical physics, but there it depended on the absolute motion of the clocks (i.e. their motion relative to the ether). Figure 3 shows that the frequency of the receding clock was decreased by the factor $(1-v/c)$ as seen from the clock at rest whereas the other way round the factor was $1/(1+v/c)$. The relativistic factor $1/f$ is seen to be the geometric mean between those two classical values. By the way, the classical formulae can still be applied, e.g. to the Doppler effect of sound in still air, if the sound velocity is called 'c'.

Let us now go back to relativity and ask for the time recorded by Ed's clock while unit time passes in the rest system. The simplest way to do that is by using two clocks at rest, say, those of Jack and Mac. Let us assume they are the distance v apart; then if Ed passes Jack at time zero he passes Mac at time 1 (by

Mac's clock). Jack watches for that event and sees it when his own clock reads $(1+v/c)$. Ed's clock seems to him slow by the factor $1/f$, so it reads $(1+v/c)/f = \sqrt{\{1-(v/c)^2\}}$ at the time it passes Jim, whose clock, as we saw, reads unit time. So here is our answer: judged by two synchronized clocks at rest, a clock moving at the speed v goes slow by the factor $\sqrt{\{1-(v/c)^2\}}$. That is the reciprocal of the 'Lorentz factor' $\gamma = \{1-(v/c)^2\}^{-\frac{1}{2}}$, the ratio of the relativistic mass and the rest mass of a body moving at speed v; $1/\gamma$ gives the 'Lorentz contraction', the factor by which a rod appears shortened when judged from a frame in which it moves (lengthwise) at speed v.

This is the 'relativistic time dilatation'. It is not in conflict with the relativity principle; I have been careful to say 'judged by *two* synchronized clocks ...'. To time a clock moving at uniform speed relative to an inertial frame we need two (or more) synchronized clocks in that frame. Thus we compare one clock with two others, a situation which is essentially unsymmetrical. We can of course ask about the rate of Jack's 'resting' clock with respect to the two clocks of Ed and Fred; relative to them, Jack's clock will now be losing, in complete agreement with the relativity principle.

It is vague and misleading to say 'a moving clock goes slow'. To be precise, one should say: 'a clock moving at speed v relative to an inertial frame containing synchronized clocks is found to go slow by the factor $1/\gamma = \{1-(v/c)^2\}^{\frac{1}{2}}$ when timed by those clocks'. That statement does not contradict the relativity principle but indeed follows from it.

There is a way of comparing a moving clock with just one clock at rest, if the moving clock changes speed and comes back to pass the other clock a second time. Let us say Albert with his clock accompanies Ed till they reach Mac; at that point Albert reverses speed and returns to Jack, taking the same time as for the outward journey. Jack will say to him 'you have been away two time units', but Albert's clock will read only $2/\gamma$. This really seems to mean that a moving clock goes slower than one at rest, and that would constitute a paradox: from the relativity principle we seem to have derived a result that contradicts it. But again our formulation was inaccurate: Albert's clock has lost time relative to Jack's, not because it has moved but because it

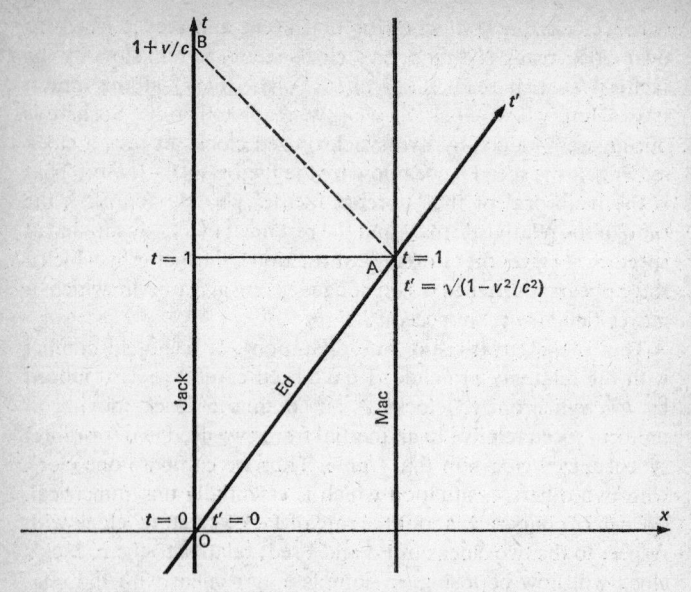

Figure 4 Relativistic time dilatation. Let Jack and Mac be the distance v apart where v is Ed's speed; then Ed, having passed Jack at zero time, reaches Mac when Mac's clock reads unit time. That event is seen by Jack at the time $1+v/c$ (event B). The reading on Ed's clock must be $(1+v/c)/f$ where

$$f = \sqrt{\{(c+v)/(c-v)\}}$$

is the Doppler factor. So Ed's clock reads

$$(1-v^2/c^2)^{\frac{1}{2}} = 1/\gamma$$

as the time difference between the two events O and A whereas that time difference is one time unit according to the clocks of Jack and Mac. Hence a moving clock appears slow by the factor $1/\gamma$ when timed on passing two synchronized clocks at rest

has changed its motion on the way. So the situation once again is unsymmetrical.

It may help to discuss an example, designed to give simple numerical relations. Albert leaves his twin brother Jack – who is said to be at rest – at Christmas at a speed of $0.8\,c$ and travels for three years, by his own clocks; he then reverses speed and gets home after another three of his years; so he would say, he has

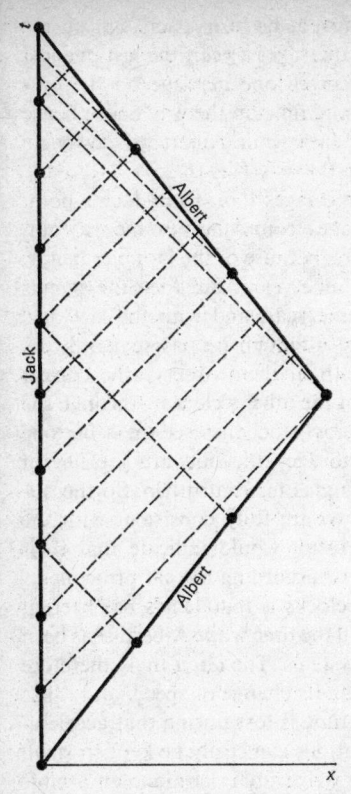

Figure 5 An example of the 'twin paradox'. Albert leaves his twin brother Jack, going for three years (by his clocks) at the speed $0·8\,c$; then he reverses speed and returns home after another three years. His Christmas greetings at first are delayed by the Doppler factor $\sqrt{\{(1+0·8)/(1-0·8)\}} = 3$; but after nine years Jack sees Albert approach, his greetings come every four months, and at the end of the tenth year he is home. He in turn received only one of Jack's greetings at the very end of his outward journey (one every three years, the same Doppler factor), and nine on the way back (three years). It all fits the expected time dilatation: $1/\gamma = \sqrt{\{1-(0·8)^2\}} = 6/10$

been away for six years in all. Both Jack and Albert send each other regular Christmas greetings by radio. The Doppler factor is $\sqrt{\{(1+0·8)/(1-0·8)\}} = 3$; hence Albert receives the first message only after three years, just as he turns back. But during the return journey he gets three messages a year, the last one just as he gets home. So he has received one message on the outward journey (at the end of it), and nine on the way back; hence ten years must have passed for Jack while Albert was away six of his years. This fits: the factor $1/\gamma = \sqrt{(1-16/25)} = \frac{3}{5}$.

How does this exchange of messages look from Jack's point of view? He receives three messages from Albert on the way out, arriving at intervals of three years, because of the Doppler factor. So he sees Albert receding for nine years. But then the signals speed up: three arrive within one year, and with the last one, Albert is home, after ten years, although he has sent only six annual messages. It all fits; both of them observe the correct Doppler slowing (or speeding) of the other's clocks – through the annual messages – depending on whether they see the other one receding or approaching. But to Jack the outward journey of Albert appears nine times as long as his return trip. So the different times recorded by the two are fully consistent with the relativity principle; any other result would indicate that their clocks had not been synchronized according to that principle.

The difference between their clocks is that Jack's has been at rest in the same inertial system all the time while Albert's has been at rest in two different inertial systems. The latter must therefore have suffered acceleration, during its change of speed, and it has been sometimes suggested that time is lost during that acceleration. But actually those accelerations can easily be kept so small that they would hardly affect a wristwatch, let alone an atomic clock. Furthermore, if the overall journey was made longer the time lost would go up in proportion with the duration of the journey, while any effect of acceleration would remain unchanged and hence become less relevant.

Let me point to a simple analogy. Two motorists go from A to C, Joe in a straight line, Bill via B. On arrival they find that Joe has travelled sixty miles and Bill seventy. Surely this is no paradox! Admittedly Joe has travelled in a straight line all the time and Bill most of the time, but no one would say that Bill acquired his extra mileage at the corner he had to turn at B. We are all familiar with the fact that a broken line is *longer* than a straight line between the same two points. But most of us are not yet familiar with the fact – which follows from the relativity principle – that the time interval between the same two events is *shorter* when measured along a broken line (i.e. by a clock that changes its speed) than when it is measured along a straight line (by a clock travelling at constant speed).

Sometimes the objection is made that in general relativity all frames of reference are equally admissible and that therefore the situation between Jack and Albert is not really unsymmetrical; that argument will be discussed in the next reading.

Some people are willing to believe that clocks behave like that, but doubt whether Albert would really return looking four years younger than his stay-at-home twin. That is indeed doubtful; he may well have aged a lot more then Jack, because of the discomforts of space travel! But an organism is a clock, though a poor one, easily affected by its surroundings; if those extraneous effects are eliminated or allowed for it must behave like any other clock if the relativity principle is valid.

There is, however, yet another way for comparing two clocks in relative motion, if we bring in another coordinate of space. Let us say, Jack wishes to compare his clock with that of the distant traveller Dan, who moves at speed v parallel to the x-axis and at distance b away from it. If b is large enough Dan will spend an appreciable time near enough to the y-axis (Figure 6) so that the signals he sends out during that time will all suffer practically the same delay in reaching Jack; then Jack can just look at Dan's clock, without any need for corrections. Now Dan is at rest relative to Ed, and we remember that Ed's clock is slow by the factor $1/\gamma = \sqrt{\{1-(v/c)^2\}}$ if timed by Jack and Mac as it passes them; so we conclude that Dan's clock must be slow by the same factor if timed in the same way. But actually that proviso is unnecessary: Dan's signals are received by Jack and Mac essentially simultaneously, provided they come from a distant point near the y-axis. So the timing may be done entirely by Jack, with the same result: that Dan's clock goes slow by the usual factor $1/\gamma$.

But is this not another paradox? Here we have two clocks each at rest in an inertial frame, synchronized according to the relativity principle; how then can one of them go slower than the other?

Let us be precise. What Jack observed is that Dan's clock

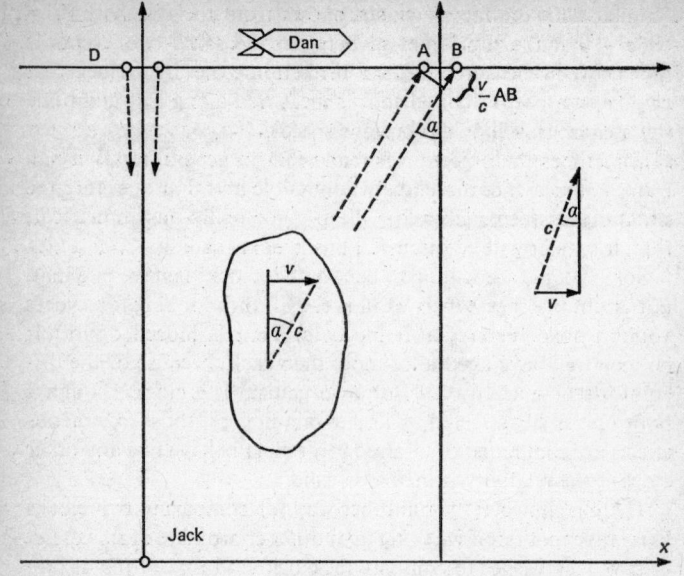

Figure 6 The 'transverse Doppler effect'. This is *not* a space–time diagram. The bold arrow DB indicates the path of the distant traveller Dan; the two dotted lines on the left the paths of two light signals, sent out one time unit apart by Dan while close to the y-axis. They will travel the same distance (in first order): Jack will see them from the y-direction and receive them $1/\gamma$ time units apart, because of time dilatation (see the text for the detailed argument). What about time signals from Jack, seen by Dan from the y-direction? They do not travel parallel to the y-direction in the rest frame but at an angle a where $\sin a = v/c$ (see insert). The time interval between their arrivals at Dan's at the points A and B is AB/v (since Dan travels at the speed $v = (1+v/c) \cdot AB$) or equal to $1/(1-v^2/c^2)$. This is longer by the expected factor γ than the time unit on Dan's clock; so Dan, looking along the y-axis sees Jack's clock slow by the factor $1/\gamma$ just as Jack sees Dan's clock

signals, received from a direction perpendicular to the *x*-axis (their line of relative motion) are slow compared to his own. The relativity principle demands that the reverse statement, obtained by exchanging the names 'Jack' and 'Dan', should also be true. And that is indeed so. The point is that the signals that Dan sees arriving from a direction parallel to the *y*-axis do not travel in that direction *relative to the rest frame*: they travel at the angle a relative to the y-axis where $\sin a = v/c$. So in the rest system the delay between two light signals in reaching Dan is not the same although it is (practically) the same if referred to Dan's system. Let us compute that difference in delay, referring everything to the rest system. Let A and B be the two points at which two light signals, emitted one time unit apart by Jack's clock, reach Dan. The second light signal has the extra distance $(v/c) \times AB$ to go, so they arrive $(1+v/c) \times AB$ time units apart, and in that time Dan has covered the distance AB, going at the speed v. Hence $(1+v/c) \times AB = AB/v$, or $AB/v = 1/\{1-(v/c)^2\}$. That is the interval between the arrivals of the two unit ticks from Jack's clock, and it is indeed longer than Dan's time unit, by the expected factor γ; so he will report that Jack's clock goes slow by the factor $1/\gamma$, as we know he must.

I hope I have made it clear that the behaviour of clocks is not really paradoxical in the cases I have discussed. This does not prove that no real paradox, no logical inconsistency, can be found anywhere in the theory of relativity; such a proof must be mathematical and has indeed been given.

You may still ask what guarantee we have that clocks really behave in a manner consistent with the relativity principle. Well, the most direct proof comes from the study of beams of unstable particles such as mesons. Such a beam constitutes a clock of a kind: the number of particles diminishes along the beam as $\exp(-t/t_0)$ where t_0 is the mean life of the particles and t is their 'proper time', i.e. the time as measured with a clock that moves with them. For instance for (charged) pions $t_0 = 2 \cdot 5 \times 10^{-8}$ s in which time a pion could travel 25 ft [7·6 m] if it went at the speed of light. Without the relativistic slowing-down of clocks, a pion beam would have faded out after a few hundred feet of travel. But with big synchrotrons one can produce pions with $v/c = 0.9999$ where the factor γ is about 70; intense pion beams can be obtained extending to many hundred feet from the target, and indeed the decrease in beam intensity along the beam confirms that the pions decay about 70 times more slowly – by laboratory time – than when they are at rest. Less clear-cut but equally striking examples have long been known from cosmic rays.

A somewhat more indirect proof was obtained by Ives and Stillwell. They accelerated hydrogen ions to about 40 keV ($v/c = 0.006$); some of the ions then captured an electron and emitted spectral lines, whose Doppler shift was accurately measured. The transverse Doppler effect could not be observed directly because it would have amounted to only 0·002 per cent change in wavelength, which would have been masked by an uncertainty of less than 0·2° in the direction (relative to the particle stream) of the light observed. Instead they measured the Doppler shifts at two directions forming the same small angle ε with the forward and the backward direction of the particle stream; the mean of the two shifts is

$$\tfrac{1}{2}[\sqrt{\{(c+v)/(c-v)\}} + \sqrt{\{(c-v)/(c+v)\}}]\cos \varepsilon = \gamma \cos \varepsilon,$$

and this value was indeed observed.

There is also an experiment by Hay, Schiffer, Cranshaw and Egelstaff who showed that gamma-ray absorption is diminished by fast transversal motion of emitter and absorber; but this experiment employed rotary motion and will consequently be discussed in a later article.

It may also be asked by what mechanism a clock adjusts its rate when its state of motion is altered. That mechanism could probably be worked out for each type of clock, but that would usually be a very difficult exercise, and not really necessary. We have no reason to doubt the accuracy of the relativity principle. Several of its consequences have been accurately tested. Many elastic collisions between fast protons and protons at rest have been seen in cloud and bubble chambers, and the speed and angles of the two protons resulting from such collisions agree with Einstein's formulae; and in proton synchrotrons the magnetic field must be made to increase accurately as a function of the accelerating frequency, a function computed from the relativistic mass increase, or else the synchrotron does not work. In view of all this evidence it seems sensible to use the relativity principle when it offers us a quick answer to a question, just as we use the principle of energy and momentum conservation to predict – in part – the behaviour of a mechanical system without laboriously integrating its equations of motion.

All the same it may be instructive to consider a very simple

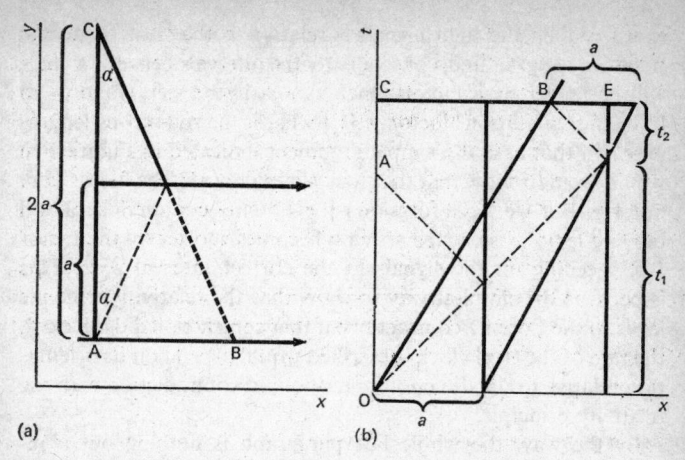

Figure 7 Time dilatation illustrated by a 'rod clock'. In (a) the bold lines represent the paths of the two ends of a rod of length a, travelling crosswise at the speed v. A light signal, travelling back and forth along the rod, traces out the dotted line, with $\sin \alpha = v/c$. Hence the time interval between the ticks A and B is

$$AC/(c \cos \alpha) = (2a/c)(1-v^2/c^2)^{-\frac{1}{2}} = 2a\gamma/c,$$

compared to $2a/c$ for the same clock at rest; this is the usual time dilatation.

(b) is a space–time diagram, representing the same rod clock in lengthwise motion, without Lorentz contraction, together with one at rest. In the latter, a light signal emitted at O returns at A, after a time $2a/c$. In the moving clock it comes back at B, which is synchronous (in the rest frame) with C. The time interval $OC = t_1+t_2$; $t_1 = OD/c = (OD-a)/v$, hence $t_1 = a/(c-v)$: $t_2 = BE/c = (a-BE)/v$, hence $t_2 = a/(c+v)$. So

$$OC = a/(c-v)+a/(c+v) = OA/(1-v^2/c^2) = OA \cdot \gamma^2.$$

Thus the moving clock seems to be slowed by the factor $1/\gamma^2$ instead of $1/\gamma$. But we must remember that a rod moving lengthwise suffers Lorentz contraction by the factor $1/\gamma$; then the time dilatation comes out correctly

type of 'clock', namely a rod of length a with a flash lamp at each end, so constructed that each lamp will be set off by the flash from the other. Thus the lamps will flash alternately, the interval between two flashes being a/c, in the rod's rest system. If the rod moves at right angles to its own direction, with the speed v, then the light signals – relative to the 'rest' system – trace out a zigzag line, consequently the intervals between signals will be longer by a factor which can easily be seen (Figure 7a) to be the 'dilatation' factor γ. But what if the rod moves lengthwise? In that case the simple argument indicated in Figure 7(b) would seem to show that the clock slows down by the factor $1/\gamma^2$, not $1/\gamma$. But we have forgotten the Lorentz contraction: a rod moving lengthwise at the speed v becomes shorter by the factor $1/\gamma$, speeding up the signals to the correct interval $a\gamma/c$. This is perhaps the simplest way to show that the relativity principle leads to the Lorentz contraction: if that contraction did not exist, the rate of the 'rod clock' described would depend on its orientation relative to the direction of motion, in contradiction with the relativity principle.

By the way, the whole last paragraph is nothing but a rewording of the traditional discussion of the Michelson–Morley experiment: the two arms of the interferometer can be considered as two 'rod clocks', and the historic experiment established their synchronous behaviour irrespective of orientation.

So we should accept that within the framework of special relativity, i.e. for clocks that for most of the time are at rest in some inertial system, with no large differences in gravitational potential, the terms 'identical clocks' and 'clocks synchronized by the relativity principle' are equivalent. That is no longer so with accelerated clocks or in the presence of strong gravitational potential differences, as we shall see in the next Reading.

15 O. R. Frisch

Time and Relativity II

O. R. Frisch, 'Time and relativity: Part II', *Contemporary Physics*, vol. 3, 1962, no. 3, pp. 194–201.

In the last Reading we have seen that clocks behave in a somewhat surprising manner according to the special theory of relativity: two events, simultaneous by the clock in one inertial frame, will in general not be simultaneous by the clocks in another inertial frame, and of two clocks in relative motion, each will appear slow to an observer moving with the other. But there was no reason to think that two identical clocks at relative rest would not go at the same rate. We shall now show that this is no longer so for clocks in an accelerated system, or in the presence of gravity.

Let us consider a spaceship of length b with its rockets going at full blast, giving it an acceleration a; Francis (in front) and Robert (at the rear) have identical clocks. Light takes a finite time b/c to go from Francis to Robert, and during that time the ship gains the speed ab/c; we shall assume that speed to be small compared with c. By the time Robert receives a signal – say a train of monochromatic light or radio waves – he has the speed ab/c relative to the inertial system in which Francis was at rest when the signal was sent out. Hence he will observe the signal with a Doppler shift, that is with a frequency increased by the factor $\sqrt{\{(c+v)/(c-v)\}} \simeq 1+v/c = 1+ab/c^2$. This factor is independent of the frequency of the light and applies equally well to the ticks of Francis' clock; hence that clock will appear fast to Robert, by the factor $1+ab/c^2$. In the same way we can see that Robert's clock will appear slow to Francis, by the factor $1-ab/c^2$. All this is a straightforward consequence of the Doppler effect and does not even depend on the principle of relativity.

Inside the ship any object released will no longer be accelerated and will consequently drop behind; it will be seen to move with the apparent acceleration a toward the tail of the ship. Einstein pointed out that this behaviour of objects in the ship might equally well be due to the gravitational attraction of a planet on which the ship was standing on its tail, provided the empirically found proportionality between weight and mass is strictly correct. He took an important step beyond this when he proposed his equivalence principle, by which a uniform gravitational field is equivalent to an accelerated frame of reference in every aspect, including the behaviour of electromagnetic field such as light signals. From this he developed later his 'general theory of relativity'; but here we shall be concerned only with the direct consequences of the equivalence principle. In particular we can see that two clocks, placed at the top and the foot of a tower, will behave just like our two clocks in the nose and the tail of an accelerating spaceship; we have only to replace b, the length of the ship, by h, the height of the tower, and a, the acceleration of the ship, by g, the acceleration due to terrestrial gravity. The equivalence principle then predicts that the observer on the top of the tower would see the other clock go slow compared to his own identical clock by the factor $1-\Delta\phi/c^2$ where $\Delta\phi = hg$ is the difference in gravitational potential; for instance excited atoms near the foot of the tower would emit spectral lines which have lower frequency (are shifted toward the red) compared to the lines emitted by the same kind of atoms at the top.

It may be well to make this result plausible in a different way. Let us consider two equal atoms at a different height, connected over a pulley by that useful piece of hardware, a weightless string (Figure 1). We assume that to begin with the lower atom is in its first excited state, possessing the extra energy E and hence – according to the special relativity theory – the extra mass E/c^2. We then allow it to emit that extra energy as a photon which travels upwards until it is absorbed by the upper atom. This will now be the heavier and will sink down, turning the pulley and doing work amounting to $hg \cdot E/c^2 = E \cdot \Delta\phi/c^2$. Now we are apparently again where we started and can repeat the process indefinitely, obtaining work all the time. Have we indeed invented perpetual motion?

Surely not; the work done by the pulley must come from somewhere. Everything fits if we assume that the photon on arriving at the upper atom has not quite enough energy to excite the upper atom; indeed its energy must fall short just by $E \cdot \Delta\phi/c^2$, the work we expect from the pulley. So the quantum must arrive with the energy $E(1-\Delta\phi/c^2)$, and if we remember the proportionality between frequency and energy of a light quantum we see that this tallies with the conclusion we got from the equivalence principle.

Actually it is not necessary to bring in the quantum theory;

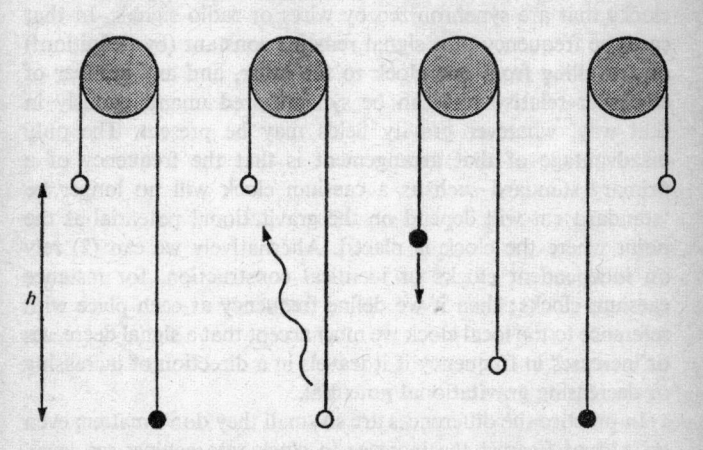

Figure 1 Two identical atoms, hung over a pulley by a weightless thread, apparently make perpetual motion possible; the puzzle is resolved by the gravitational red-shift of photons (see text)

merely from the Lorentz transformation of energy and momentum it can be shown that the energy contained in a light signal changes by the same factor as its frequency if it is observed in a different frame of reference. So the proportionality in which our agreement rests need not be taken from the quantum theory; it is implicit in the special theory of relativity. The fact that the atoms suspended from our weightless string have quantized energy states is not really important; we could have used searchlights instead and would still have to come to the conclusion that a light signal on climbing against gravity loses energy and hence frequency.

But, you will ask, how can the signal lose frequency? Since the distance between source and receiver is constant, surely the number of wave crests that arrive at the receiver must equal the number that left the source during the same time; so the frequency must be the same! That is indeed so provided we use clocks that have been synchronized by signals, say from a common master clock, rather than clocks of identical construction, as we have assumed so far. In the presence of gravity we must distinguish between two ways of establishing time units in places of different gravitational potential. We can (1) set up a system of clocks that are synchronized by wires or radio signals. In that case the frequency of a signal remains constant (by definition!) on travelling from one clock to the other, and any number of clocks at relative rest can be synchronized unambiguously in that way, whatever gravity fields may be present. The only disadvantage of that arrangement is that the frequency of a primary standard such as a caesium clock will no longer be 'standard': it will depend on the gravitational potential at the point where the clock is placed. Alternatively we can (2) rely on independent clocks of identical construction, for instance caesium clocks; then if we define frequency at each place with reference to the local clock we must accept that a signal decreases or increases in frequency if it travels in a direction of increasing or decreasing gravitational potential.

In practice the differences are so small they don't matter; even on Mount Everest the increase in clock rate against sea level, hg/c^2, would be only one part in 10^{12}. However, caesium clocks and other atomic clocks are now approaching that accuracy, and an experimental test may soon be possible. A much greater difference in potential, $\Delta\phi/c^2 = 2 \cdot 12 \times 10^{-6}$, exists between the Earth and the surface of the sun, and even higher potentials must exist on the surface of the very dense white dwarf stars. Einstein predicted as early as 1909 that spectral lines from those sources should be shifted to the red, compared with the lines of the same atoms on Earth. This gravitational red-shift has been looked for most carefully but is obscured by a variety of other line shifts, and although there is some evidence for it the results don't seem to be conclusive.

However, a recent discovery by Mössbauer has made it possible to use certain radioactive nuclei as extremely accurate 'clocks'. It had long been realized that the frequency of the

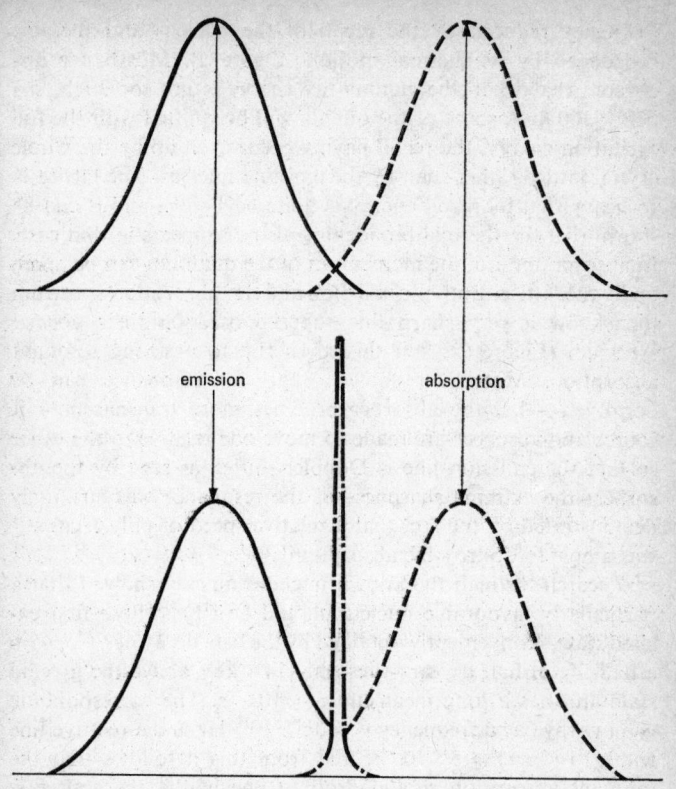

Figure 2 Both in emitting and in absorbing a gamma quantum of energy E, a free nucleus (mass m) takes up a momentum E/c and hence an energy $E^2/2mc^2$; furthermore, the thermal motion causes Doppler broadening of the line. Hence the emission and absorption lines are separated by E^2/mc^2, and both are broadened. But with bound atoms the lattice can occasionally take up the momentum and the energy lost is then negligible; there appears an unshifted line, both in emission and absorption, causing strong resonance over a width, determined chiefly by the mean life of the excited state

gamma rays emitted from nuclei is extremely well defined; for instance with a frequency v around 10^{20} Hz and a mean life time τ of 10^{-7} s the natural line width $\Delta v/v = 1/2\pi v\tau$ would be less than 10^{-13}. But the gamma rays actually observed have their frequency reduced by the recoil of the nucleus and the line broadened by its thermal motion (Figure 2). Mössbauer discovered that when the gamma-ray energy is not too high, say below 100 keV, some of the quanta will be emitted with the full excitation energy, the recoil having been taken up by the whole crystal lattice rather than by the emitting nucleus. The lattice is so heavy that its recoil energy is quite negligible, and it can be shown that the thermal broadening also disappears in that case. In absorption, too, the momentum of the quantum can be taken up by the lattice. Both the emission and the absorption spectrum then show a very sharp line superimposed on their normal spectrum (Figure 2), and this gives rise to a strong resonant absorption. Mössbauer showed that this resonance can be destroyed – i.e. the absorber becomes more transparent – if source and absorber are made to move one relative to the other so that the emission line is Doppler-shifted as seen by the absorber; the extreme sharpness of the resonance was strikingly demonstrated by the fact that a relative speed of only 1 cm s^{-1} was ample to destroy the absorption!

A search through the known nuclides quickly showed that a particularly favourable nucleus should be ^{57}Fe, whose first excited state – conveniently obtained by the beta decay of ^{57}Co, with a half-life of half a year – lies only 14·4 keV above the ground state and has a long mean life $\tau = 10^{-7}$ s. The corresponding gamma ray has a frequency $\nu = 3 \cdot 5 \times 10^{18}$ Hz and a relative line width $1/(2\pi\nu\tau) = 5 \times 10^{-13}$, and from this it follows that the resonant absorption should drop to one half if the source is moved relative to the absorber at a speed of about 0·01 cm s^{-1}. Experiments (Figure 3) showed the line to be slightly wider depending on the material used and its heat treatment. The reason is probably that the line is actually a Zeeman multiplet formed by the effect of the atomic magnetic field on the nucleus, and it is difficult to make the fields exactly alike in source and absorber.

Even so, here was clearly a possibility to detect extremely small frequency changes, and the idea of looking for Einstein's gravitational shift occurred to a number of people. All that seemed necessary was to mount the source and the absorber, one above the other, as far apart as the gamma-ray intensity would allow, and then look for a small asymmetry in the Doppler-shift curve

(Figure 3). A race developed between Harvard and Harwell, in which Harwell obtained the first indication of a shift (Cranshaw, Schiffer and Whitehead); but doubts were cast on that result when it was found by Pound and Rebka that a shift of the same magnitude might have been produced by a temperature difference of less than 1 °C between source and absorber.

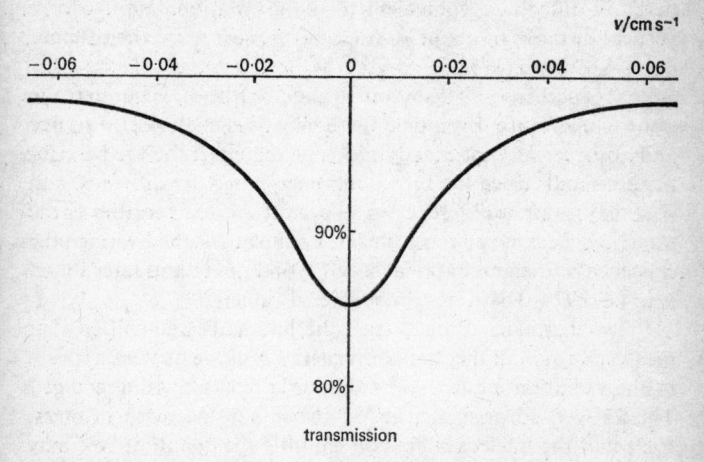

Figure 3 Typical transmission, through a thin iron foil, of the 14·4 keV gamma rays of excited ^{57}Fe, as a function of the relative velocity of source and foil. The absorption is almost entirely due to the 'recoilless' resonance, which is so narrow that the Doppler effect from quite slow relative motion is enough to spoil it

Such a temperature effect had independently been foreseen by Josephson and can be described simply as the transverse (quadratic) Doppler effect due to the thermal motion of the nuclei. The ordinary, linear Doppler effect is of course many thousand times larger, but is absent in the Mössbauer line; one may say that it is cancelled out because the emitting nucleus, carrying out its thermal oscillations, changes its direction many million times during the mean life of the excited state. But there remains the second-order Doppler effect, or in other words the relativistic dilatation, discussed in the previous reading. This amounts to $\Delta v/v = \frac{1}{2}v^2/c^2 = U/2c^2$ if we assume that $\frac{1}{2}v^2$, the mean kinetic energy per unit mass of the lattice, is half its total energy U per unit mass, as it will be if the nucleus is bound to the lattice by elastic forces.

Of course if the absorber is at the same temperature as the source one expects no observable shift, but if they differ by ΔT the shift will be $\Delta T \cdot C_p/2c^2$ where C_p is the specific heat. For iron at room temperature this comes to $2 \cdot 2 \times 10^{-15}$ for 1 °C temperature difference, equivalent to the gravitational shift over a vertical distance of about 70 ft [21 m], almost twice the distance used by Cranshaw *et al.*

So it is necessary to keep source and absorber at accurately the same temperature. Even then there may be a small shift if source and absorber are not exactly the same material: the Debye temperature and hence the zero point energy may be different, and that may cause a difference in U even if the temperature is the same. By checking all those effects, Pound and Rebka verified the existence of the gravitational shift beyond doubt and later found it to be 0.97 ± 0.04 times the predicted value.

If the thermal motion caused the line to be red-shifted that means that not all the excitation energy E of the nucleus appears in the radiation emitted; what happens to the missing energy? The answer, surprisingly, lies in the minute decrease in mass, E/c^2, that the nucleus suffers on emitting the quantum. We may consider the nucleus as bound by harmonic forces to its place in the lattice; those forces remain unchanged, and hence its oscillation frequency f – which is proportional to $m^{-1/2}$ – will increase by $\Delta f/f = \frac{1}{2}\Delta m/m = \frac{1}{2}E/mc^2$. The fact that the Mössbauer effect is recoilless means that the oscillation quantum number remains unchanged, so the oscillation energy $Q = Um$ of the atom will increase in proportion with the frequency and thus by the amount $\Delta Q = Q \cdot \frac{1}{2}E/mc^2 = E \cdot U/2c^2$, just the amount that we saw was missing. It may be objected that individual atoms do not behave like separate oscillators, but form part of a lattice of many atoms, and that the effect of the change in mass is therefore many times smaller; but then we must also remember that many different oscillation modes of the lattice are affected, and if this is properly allowed for one gets the same answer as before.

It is perhaps instructive to go back to the transverse Doppler effect, discussed in the previous article, and ask what happens to the energy there. Consider an excited atom, with the excitation energy E, travelling at the speed v relative to our 'rest system'. If it emits its energy as a quantum that, in the rest system, travels at right angles to the motion of the atom, that quantum will be red-shifted and will deliver only the energy $E\sqrt{(1-v^2/c^2)}$ to an absorber at rest. If we ignore powers higher than v^2/c^2 an energy amount $E \cdot \frac{1}{2}v^2/c^2$ is missing. What has happened to the missing energy? Here again the clue is that the atom has become lighter by $\Delta m = E/c^2$. In the rest system its momentum will be unchanged since the photon went off at right angles (ignoring, for the moment, the small recoil effect); so its kinetic energy $E_k = p^2/2m$ has increased by $\Delta m \cdot d(p^2/2m)/dm = E \cdot \frac{1}{2}v^2/c^2$ which is just the missing energy.

Actually with a little algebra we can solve the problem exactly for any speed of the atom, and allowing for the recoil which causes it to change direction. We find that the energy of the emitted photon is changed by the factor $(1-E/2mc^2)\sqrt{(1-v^2/c^2)}$. The second term in the first bracket disappears when we neglect the mass E/c^2 of the photon compared to that of the 'clock', as we usually do in relativity theory. In the same way we can verify that the law of energy conservation holds for the ordinary, linear Doppler effect; but in that case the recoil of the clock, even if it is treated as very heavy, remains significant.

Let us come back to the temperature effect of the Mössbauer line. The measurements have shown it to be in good agreement with special relativity, and this also shows that those nuclear clocks are wonderfully insensitive to shaking; their thermal motion subjects them to incessant accelerations of about $10^{16}g$ ($\simeq 10^{19}$ cm s^{-2}) without affecting their rate by more than one part in 10^{13}.

A more direct test of the effect of speed on clock rate was performed by Hay, Schiffer, Cranshaw and Egelstaff by having the source near the centre and the absorber at the periphery of a fast spinning wheel. Speeds up to about 200 m s^{-1} were used, giving an expected time dilatation by about two parts in 10^{13}, nearly one half of the line width for ^{57}Fe. The difficulty here was to avoid the linear Doppler effect, which in the line of motion would have been about a million times larger. It would be impossible to mount the source so close to the centre of the wheel that the absorber would have no significant radial velocity component relative to the source. The problem was solved by attaching the source to the wheel, near its centre; then as long as the wheel is rigid enough (and this was checked) the distance between any parts of the absorber and of the source remains constant despite the rotation, and thus there is no first-order Doppler effect. Figure 3 shows that for small speeds the increase in transmission goes with the square of the line shift, which in turn goes with the square of the speed. The resulting increase of transmission with the fourth power of the speed was indeed observed, up to the calculated increase by about 4 per cent at 200 m s^{-1}. Of course the direction of the line shift could not be verified in this experiment, only the amount.

It is quite legitimate to use special relativity as long as we refer all motions to an inertial system, most conveniently to that in which the centre of the wheel is at rest. But what if we use a frame of reference that rotates with the wheel? In that frame both source and absorber are at rest; but an observer experiences a centrifugal acceleration $r\omega^2$ which he attributes to a gravitational force with the potential $\frac{1}{2}r^2\omega^2 = \frac{1}{2}v^2$. Consequently if the observer is at the edge of the wheel he expects photons, emitted near the centre, to be blue-shifted by $\Delta v/v = \phi/c^2 = \frac{1}{2}v^2/c^2$ on arriving at the periphery. But this is just what we deduced from special relativity, using an inertial frame of reference. The experiment of Hay *et al.* has sometimes been said to confirm a conclusion from general relativity; but actually special relativity is quite sufficient to predict its outcome, and the equivalence principle need not be invoked.

This is perhaps a good place to go back to the twin paradox, as discussed in the previous reading. There we concluded that Jack, who has been at rest in the same inertial frame all the time, experienced the passage of a longer time than his twin brother Albert who travelled out into space and back again. But are we not entitled to use Albert's ship as our frame of reference? In that frame, Albert is stationary throughout while it is Jack who moves out into space and back again. So now it looks as if it should be Albert, not Jack, whose clock shows the passage of a longer time. Since those two answers are in conflict it has sometimes been argued that neither can be true, and that both brothers must age the same, whichever way they move.

But if we choose a frame in which Albert is at rest and Jack goes out and back again, then in that frame there must be a gravitational field to account for the accelerations experienced by Albert, and for the fact that Jack does not feel any acceleration although he goes out and back. We shall show that if we include the frequency shifts produced by those gravitational fields we arrive at the same conclusion as before, and that there is no conflict between the two treatments.

To simplify the argument we shall assume again that the speeds are so small that powers higher than $(v/c)^2$ can be neglected. The initial speed-up and the final slow-down are unimportant because Jack and Albert are so close together that they are both at almost the same gravitational potential (and might indeed avoid those accelerations by comparing their clocks in flight as they pass each other without matching speed, neither initially nor at the end). The important period is that during which Jack – in Albert's system! – reverses speed. Let us say that Albert puts on his rockets for the time t so as to produce an acceleration a; then in his frame of reference there is a corresponding gravitational field that counteracts the rockets and, during the time t, gives the acceleration a to Jack. The distance between the brothers at that period is vT where T is the time they have been flying apart at the speed v. (In our approximation it does not matter whether T is measured by Jack or Albert.) Therefore the gravitational potential difference between them is avT, and the gravitational shift causes Jack's clock to gain the amount $tavT/c^2$ during the time t the acceleration is on. Now t has to be long enough to cause reversal, that is $at = 2v$. Hence the time gained by Jack's clock is $2Tv^2/c^2$. But this is just twice the amount which it should have lost according to the argument from the previous Reading; remember that in Albert's frame, which we have now been using, it is Jack who has been the traveller. Altogether Jack's clock has gained Tv^2/c^2 and this is just the amount we found previously by referring everything to Jack's inertial system. So the answer is the same whichever of the two frames of reference we use.

Of course we have demonstrated this only for small values of

v/c where powers higher than $(v/c)^2$ can be neglected. Otherwise we would have needed the mathematical apparatus of general relativity; that demonstration has been published by Born and Biem.

It remains to say a little about satellites. Circling the Earth once in ninety minutes, a low-orbit satellite has a speed of $7 \cdot 7 \times 10^5$ cm s^{-1} and hence the time dilatation amounts to $3 \cdot 3$ parts in 10^{10}, an amount easily measurable with present-day atomic clocks. On the other hand, the gravitational field between us and the altitude at which the satellite travels speeds up its clock rate. If we confine ourselves to circular orbits of radius r while r_0 is the radius of the Earth, both the velocity v and the gravitational potential ϕ are constant over the orbit; the relative speeding-up of the satellite clock compared to an identical clock stationary at sea level is then

$$\frac{\Delta v}{v} = \frac{\Delta \phi - \tfrac{1}{2} v^2}{c^2}.$$

Now
$$\Delta \phi = \int_r^{r_0} \frac{gr_0^2}{r^2} dr = gr_0^2 \left[\frac{1}{r_0} - \frac{1}{r} \right].$$

To get v we write the centripetal acceleration

$$\frac{v^2}{r} = \frac{gr_0^2}{r^2},$$

hence
$$v^2 = \frac{gr_0^2}{r}.$$

So we get

$$\frac{\Delta v}{v} = \frac{gr_0}{c^2} \left[1 - \frac{3r_0}{2r} \right] = 7 \times 10^{-10} \left[1 - \frac{3r_0}{2r} \right].$$

For low altitudes the time dilatation (the second term) prevails; above an altitude of $\tfrac{1}{2} r_0$ (just about 2000 miles [3200 km]) $\Delta v/v$ becomes positive because the first term, the relativistic blue-shift, becomes dominant. These predictions will presumably be tested within the next few years though there would seem to be little doubt about the outcome.

Further Reading

There are not many books which bridge the gap between advanced level and university degree standard, so I have also included some introductory university texts which have chapters that can be followed at this stage.

I suggest as a start books that give a general introduction to modern physics. *Modern Physics* by Smith was written especially for this level and is very readable; so are *Modern Physics* by Caro, McDonell and Spicer and *Introduction to Atomic Physics* by Tolansky. More difficult are *Atomic Physics* by Born, and *Introduction to Modern Physics* by Richtmeyer, Kennard and Lauritsen, but they provide additional information about particular topics.

For Part One a good book to start with is *The Strange Story of the Quantum* by Hoffmann. The specialist books are obvious from their titles.

Nuclear Physics receives extensive treatment in the general books I have mentioned earlier, but these do not have so much about fundamental particles; for this topic try *Elementary Particles* by Frisch and Thorndike.

The books on low-temperature physics have been included because they contain chapters on superconductivity, giving some of the background for Part Three. *Gases, Liquids and Solids* by Tabor, and *Semiconductors* by Wright also provide reading in solid-state physics.

Books on plasma physics are rather too difficult for this stage but I have included one or two in the list.

For Part Five I strongly recommend *Relativity and Common Sense* by Bondi. Bondi's approach to relativity has been widely acclaimed as providing a very good introduction to the subject.

D. Bohm, *Quantum Theory*, Prentice Hall, 1951.

F. I. Boley, *Plasmas: Laboratory and Cosmic*, Van Nostrand, 1966.

H. Bondi, *Relativity and Common Sense*, Heinemann, 1965.

M. Born, *Atomic Physics*, Blackie, 7th edn, 1962.

D. M. Brink, *Nuclear Forces*, Pergamon, 1965.

R. Brown, *Lasers*, Aldus, 1968.

D. E. Caro, J. A. McDonnell and B. M. Spicer, *Modern Physics*, Edward Arnold, 1962.

C. S. Cook, *Structure of Atomic Nuclei*, Van Nostrand, 1964.

R. P. Feynman, R. B. Leighton and M. Sands, *The Feynman Lectures on Physics*, Addison Wesley, 1963–5.

D. M. Frisch and A. M. Thorndike, *Elementary Particles*, Van Nostrand, 1964.

O. R. Frisch, *Atomic Physics Today*, Oliver & Boyd, 1962.

R. Gourian, *Particles and Accelerators*, Weidenfeld & Nicolson, 1967.

W. Heitler, *Elementary Wave Mechanics*, Clarendon Press, 2nd edn, 1956.

B. Hoffman, *The Strange Story of the Quantum*, 2nd edn, Dover, 1959 (Penguin, 1963).

D. J. Hughes, *Neutron Story*, Heinemann, 1954.

L. C. Jackson, *Low Temperature Physics*, Methuen, 5th edn, 1962.

R. Katz, *An Introduction to the Special Theory of Relativity*, Van Nostrand, 1965.

N. C. Little, *Magnetohydrodynamics*, Van Nostrand, 1968.

W. H. McCrea, *Relativity Physics*, Methuen, 4th edn, 1954.

N. F. Mott, *Elements of Wave Mechanics*, Cambridge University Press, 1962.

F. K. Ritchmeyer, E. H. Kennard and T. Lauritsen, *Introduction to Modern Physics*, McGraw-Hill, 1969.

E. M. Rogers, *Physics for the Inquiring Mind*, Princeton University Press, 1960.

M. S. Smith, *Modern Physics*, Longmans, 1960.

D. Tabor, *Gases, Liquids and Solids*, Penguin, 1969.

S. Tolansky, *Introduction to Atomic Physics*, Longmans, 4th edn, 1956.

D. A. Wright, *Semiconductors*, Methuen, 1966.

J. Yarwood, *Atomic Physics*, University Tutorial Press, 2nd edn, 1963.

M. W. Zemansky, *Temperatures Very Low and Very High*, Van Nostrand, 1965.

Acknowledgements

Permission to reproduce the Readings in this volume is acknowledged to the following sources:

1 *Contemporary Physics*
2 *Contemporary Physics* and O. R. Frisch
3 *New Scientist*
4 *School Science Review* and A. Bairsto
5 *Contemporary Physics* and R. J. Blin-Stoyle
6 The Institute of Physics and The Physical Society, and F. R. Stannard
7 The Institute of Physics and The Physical Society, and F. R. Stannard
8 *Science Journal*
9 American Institute of Physics, and J. M. Ziman
10 *Contemporary Physics* and A. C. Rose-Innes
11 *Contemporary Physics* and Alan Cotterell
12 *Contemporary Physics* and M. F. Hoyaux
13 *School Science Review* and M. M. Woolfson
14 *Contemporary Physics* and O. R. Frisch
15 *Contemporary Physics* and O. R. Frisch

Index

Absorption of light by atoms, 51
Accelerators, 107
Alfven wave, 241
Angular momentum, 77, 87, 95
Annealing twins, 206
Antimatter, 126
Atomic number, 80

Balmer series, 12
Baryon number, 129
Binding energy, nuclear, 88
Bohr atom, 21
Boson, 150
Bremsstrahlung, 27, 223
Brewster angle, 71
Bubble chamber, 117

Centre-of-mass system, energy in, 111, 138
Cloud chamber, 115
Coherence of light, 49, 61
Cohesion in metals, 202
Collective model of nucleus, 100
Compton effect, 12, 19
Correspondence principle, 23
Cosmic rays, 106, 138, 153
Critical current, 169
Critical magnetic field, 171
Cyclotron radiation, 223

Debye screening length, 219
Dislocations, 186, 209
Doppler effect, 53, 286, 305

Electric quadrupole moment, 82
Electron
  diffraction, 14, 30
  energy levels, 13, 21, 23, 50, 221
  number, 129
  spin, 14, 26, 77
Electrons in metals, 192
Electronvolt, 107

Emission of radiation
  spontaneous, 52
  stimulated, 52
Energy
  bands, 199
  levels, electronic, 13, 21, 23, 50, 221
  levels, nuclear, 84, 93
Equivalence of mass and energy, 105, 273
Exchange forces, 121
Exclusion principle, 26, 91, 99

Face-centred cubic crystal, 205
Fermi
  energy, 204
  level, 204
Frames of reference, 257, 282
Fundamental particles, 120

Gas laser, 67
Gravitational fields and relativity, 274, 282, 300

Heisenberg uncertainty principle, 30, 42, 86, 193
Holography, 15
Hyperon, 123

Incoherence of light, 53
Inertial frame of reference, 257, 282
Inverse beta decay, 138
Inverted population, 57
Ionization
  thermal, 223
  photo-, 224
  step, 228
Ionosphere, 248
Isotopes, 80
Isotopic spin, 132

Kaon, 137

Laser, 49
 gas, 67
 ruby, 59
Leptons, 137
Lorentz
 contraction, 266, 298
 force, 184

Magic numbers, 85
Magnetic
 horn, 141
 moment, 82, 96, 232
 pressure, 236
 quantum number, 77
 stars, 248
Magneto-acoustic wave, 241
Mass
 and energy, equivalence of, 105, 273
 number, 79
 variation of with velocity, 270
Meson, 122
 π-, 123, 137, 265, 295
 K-, 123, 137
 μ-, 123, 137
Metals
 cohesion in, 202
 electrons in, 192
Metastable states, 226
Michelson–Morley experiment, 261
Mobility, 232
Mössbauer effect, 302
Muon, 123, 137
 number, 129

Neutrino, 85, 87, 123, 137
Neutron, 87
Nuclear
 binding energy, 88
 collective model, 100
 emulsion, 119, 121
 forces, 89, 121

 mass number, 79
 radius, 80, 89
 shell model, 91
Nucleus, 79

Optical pumping, 57

Parity, 96
Particle
 accelerators, 107
 detectors, 114
Pauli exclusion principle, 26, 91, 99
Penetration depth, 171
Phonons, 157
Photoelectric effect, 12
Photo-excitation, 225
Photo-ionization, 224
Photons, 11, 19, 33, 44, 50, 224
Pions, 123, 137, 265, 295
Planck's constant, 11
Plasma
 definition, 218
 generators, 243
 lens, 142
 motors, 243
 waves, 240
Positron, 127
Potential well, 93
Principal quantum number, 76, 195
Proton synchrotron, 109

Q-switching of lasers, 64
Quantum
 mechanics, 27
 numbers, 22, 24, 76, 91, 195
 theory, 11, 17
Quark, 78, 134

Radiation
 cyclotron, 223
 synchrotron, 248

Radioactive decay, 82
Red shift, 275, 300
Relativity
  general theory, 299
  special theory, 253, 281
Ruby laser, 59

Schmidt diagrams, 97, 98
Schrödinger equation, 28
Screening length, 219
Semiconductors, 196
Shells, 77, 195
Shell model of nucleus, 91
Slip process, 208
Solid state, 157
Spark chamber, 118, 140
Spectra
  line, 12, 20
    broadening of, 53
Spin,
  electronic, 14, 26, 77
  isotopic, 132
  nuclear, 81, 95
Spin–orbit coupling, 89
Stacking faults, 206
Step ionization, 228
Strangeness number, 131
Sun, 247
Superconductors, 168
Surface energy, 175
Synchrotron radiation, 223, 248

Thermal ionization, 223
Transition temperature, 169
Twin paradox, 253, 266, 278, 293

Uncertainty principle, 30, 42, 86, 193

Zeeman effect, 14, 25